COMPETENCY

Philosophy and Medicine

VOLUME 39

Editors

H. Tristram Engelhardt, Jr., *The Center for Ethics, Medicine and Public Issues, Baylor College of Medicine, Houston, Texas*

Stuart F. Spicker, *School of Medicine, University of Connecticut Health Center, Farmington, Connecticut*

Editorial Board

George J. Agich, *School of Medicine, Southern Illinois University, Springfield, Illinois*

Edmund Erde, *University of Medicine and Dentistry of New Jersey, Camden, New Jersey*

Patricia A. King, J.D., *Georgetown University Law Center, Washington, D.C.*

E. Haavi Morreim, *Department of Human Values and Ethics, College of Medicine, University of Tennessee, Memphis, Tennessee*

Kevin W. Wildes, S.J., *The Center for Ethics, Medicine and Public Issues, Baylor College of Medicine, Houston, Texas*

The titles published in this series are listed at the end of this volume.

COMPETENCY

A Study of Informal Competency Determinations in Primary Care

Edited by

MARY ANN GARDELL CUTTER

Dept. of Philosophy, University of Colorado at Colorado Springs

and

EARL E. SHELP

Foundation for Interfaith Research and Ministry

KLUWER ACADEMIC PUBLISHERS
DORDRECHT / BOSTON / LONDON

Library of Congress Cataloging-in-Publication Data

Competency : a study of informal competency determinations in primary
  care / edited by Mary Ann Gardell Cutter, Earl E. Shelp.
       p.   cm. -- (Philosophy and medicine ; v. 39)
    Includes index.
    ISBN 0-7923-1304-6 (HB : alk. paper)
    1. Clinical competence.   I. Cutter, Mary Ann Gardell.   II. Shelp,
  Earl E., 1947-   . III. Series.
    [DNLM: 1. Clinical Competence. 2. Primary Health Care.   W3 PH609
  v. 39 / W 21 C727]
  RA399.A1C66   1991
  174'.2--dc20
  DNLM/DLC
  for Library of Congress   91-20800

ISBN 0-7923-1304-6

Published by Kluwer Academic Publishers,
P.O. Box 17, 3300 AA Dordrecht, The Netherlands.

Kluwer Academic Publishers incorporates
the publishing programmes of
D. Reidel, Martinus Nijhoff, Dr W. Junk and MTP Press.

Sold and distributed in the U.S.A. and Canada
by Kluwer Academic Publishers,
101 Philip Drive, Norwell, MA 02061, U.S.A.

In all other countries, sold and distributed
by Kluwer Academic Publishers Group,
P.O. Box 322, 3300 AH Dordrecht, The Netherlands.

Printed on acid-free paper

All Rights Reserved
© 1991 by Kluwer Academic Publishers
No part of the material protected by this copyright notice may be reproduced or
utilized in any form or by any means, electronic or mechanical,
including photocopying, recording or by any information storage and
retrieval system, without written permission from
the copyright owner.

Printed in The Netherlands

TABLE OF CONTENTS

PREFACE	vii
MARY ANN GARDELL CUTTER / Introduction	ix

PART I / THE CLINICAL CONTEXT

JAMES A. KNIGHT / Judging Competence: When the Psychiatrist Need, or Need Not, Be Involved	3
EDMUND D. PELLEGRINO / Informal Judgments of Competence and Incompetence	29

PART II / REEXAMINING CONCEPTS OF COMPETENCE

TOM L. BEAUCHAMP / Competence	49
EDMUND L. PINCOFFS / Judgments of Incompetence and Their Moral Presuppositions	79

PART III / FASHIONING LAW AND PUBLIC POLICY

E. HAAVI MORREIM / Competence: At the Intersection of Law, Medicine, and Philosophy	93
JOHN A. ROBERTSON / The Geography of Competency	127
PATRICIA D. WHITE and SUSAN HANKIN DENISE / Medical Treatment Decisions and Competency in the Eyes of the Law: A Brief Survey	149

PART IV / CASE STUDIES: SEEKING INSIGHT FROM APPLICATION

EUGENE V. BOISAUBIN / Competency Judgments: Case Studies From the Internist's Perspective	167
MARK PERL / Competency Judgments: Case Studies from the Psychiatrist's Perspective	179

EDWIN R. DUBOSE, JR. and EARL E. SHELP / Competency Judgments: Case Studies in Moral Perspective 185

PART V / COMMENTARY AND CRITIQUE: ANOTHER LOOK AT CONCEPTS OF COMPETENCE

HAROLD Y. VANDERPOOL / The Competency of Definitions of Competency 197
VIRGINIA ABERNETHY / Judgments About Patient Competence: Cultural and Economic Antecedents 211
STEPHEN WEAR / Patient Freedom and Competence in Health Care 227
EDMUND L. ERDE / Breaking up the Shell Game of Consequentialism: Incompetence — Concept and Ethics 237
KENNETH F. SCHAFFNER / Competency: A Triaxial Concept 253

NOTES ON CONTRIBUTORS 283
INDEX 285

PREFACE

Some conferences produce proceedings, others an inspiration to labor, which finally leads to a published work. Such has been the case with regard to this volume. In 1984, the Center for Ethics, Medicine, and Public Issues held a conference with the title 'When are Competent Patients Incompetent?' with the support of the Texas Committee for the Humanities, a state-based program of the National Endowment for the Humanities. Assistance was provided by both Baylor College of Medicine and the Institute of Religion. This conference evoked a considerable interest in examining further the moral status of competency determinations in the clinical setting. This interest is realized in this volume, which now affords us an opportunity to thank all those individuals who made the conference possible, only some of whom are acknowledged in this Preface. In particular, we wish to express our gratitude to Baruch A. Brody, Rebecca Dresser, the Honorable Jerome Jones, H. Steven Moffic, Margery W. Shaw, Eleanor Tinsley, and Albert Van Helden.

The volume took its shape through the labors of Earl Shelp and Mary Ann Gardell Cutter, who inspired the further evolution of the papers presented at the conference and attracted contributions from individuals who had not attended. Earl Shelp and Mary Ann Gardell Cutter have produced a volume following extensive reflection and dialogue; they were ably assisted in the final preparation of the manuscript by Thomas J. Bole III and George Khushf, to whom special thanks are due.

March 6, 1991
H. TRISTRAM ENGELHARDT, JR.
STUART F. SPICKER

MARY ANN GARDELL CUTTER

INTRODUCTION[1]

This volume has as its goal the examination of the various informal practices through which patients, who have not been formally or legally judged to be incompetent, are in fact treated as if they were incompetent in various areas and to various degrees. This collection brings together scholars from philosophy and law interested in the moral and philosophical presumptions of these practices, as well as physicians and psychiatrists who have reflected on these practices as they exist in health care. The volume examines the nature of competence, especially in surroundings distorted by disease and illness, the role of physicians as guardians of the best interests of their patients, and the circumstances under which individuals are no longer able to choose rationally. The contributors to this volume agree that the criteria to be used in evaluating competence are complex, that patient competence plays a key role in medical decision-making, and that competence is a graded and regional phenomena, so that patients can be more or less competent in different areas of their lives.

Major social, economic, and moral issues are at stake in informal judgments of patient competency or incompetency. In terms of such judgments, families and physicians choose with regard to who will decide about the best usage of family resources and the time of physicians and nurses. Informal determinations of incompetency may empower individuals appointed through advance directives establishing proxy or surrogate decision-makers. Since we are all potentially patients who may be regarded within such practices, much is at stake for each of us, as well as for society and families, in being clear about how such practices should best be fashioned, and why. These are central moral and philosophical issues with immediate and significant public policy implications.

In a time of increasing malpractice litigation, it is refreshing to find an area where physicians, patients, and families work outside of formal legal guidelines, and perhaps often in contravention of strict legal formalities. Patients are simply found by physicians to be incompetent, and physicians turn to next-of-kin to make decisions. Usually there is no formal court approval or oversight. Indeed, there is rarely any oversight by an ethics committee or other third parties. Social structures and expectations are evidently sufficiently intact to allow this to take place in the absence of legal sanctions and perhaps in violation of the strict requirements of the law. That things function as well as they seem without formal social intervention is a matter for some celebration.

The reader should be on notice that this volume explores an area of evolving legal and social expectations. The volume cannot provide legal advice, nor is the provision of such advice the editors' intention (in this regard, there is no substitute for the reader directly subvening an attorney). Instead, the hope is to outline a set of important moral concerns, which are at tension in the determination of competency in the context of health care. On the one hand, individuals are regarded as having the authority to defeat by their free choice the usual moral obligations of forbearance that others have to them, and to have the authority to make agreements for common undertakings. On the other hand, individuals are usually regarded as being authorities about their own best interests and life projects, including their liberty interests and how these interests may best be achieved in their own life plans. Since many individuals are in fact not the best judges of their own best interests and will choose imprudently and intemperately, the concern to respect individuals as competent authorizers of actions regarding themselves will conflict with the concern to respect individuals as competent judges of their own best interests. The first concern will be satisfied if individuals generally understand, appreciate, and are willing to accept the risks involved in their choices. The second will be satisfied only if individuals are properly informed and are helped to overcome momentary inclinations of recklessness or impetuousness. The second, in short, gives grounds for forcing patients to act in accord with their long-term interests and life plans, though considerations regarding their long-term liberty interests will set some limitations to such paternalistic interventions. Still, there will often be an argument on behalf of temporarily restricting capricious free choice in order to preserve long-range liberty interests, that is, to force people to be free.

Even if one examines competence primarily to determine when patients are moral agents capable of authorizing their involvement in or refusal of treatment, there will be concerns regarding false-positive determinations of competence. The more the downside risks associated with a false positive determination increase, the more one will wish to make sure that a patient is in fact choosing competently. Consequently, when a patient for bizarre or idiosyncratic reasons refuses a standard lifesaving treatment, which would be able to restore a patient to health, one will have good moral grounds to make quite certain that the patient is not psychotic or otherwise incompetent. A bizarre choice is always a possible indication of incompetence. The costs associated with a false-positive determination in circumstances in which treatment to be refused would be life-threatening are considerable. On the other hand, when a patient appears competent and chooses an accepted or standard form of treatment, one has very little ground for expending scarce time and energies in making sure that the patient is in fact competent. If a patient who is terminal and whose death is imminent requests some relatively harmless but bizarre treatment and who appears competent, physicians may acquiesce to his wishes thereby providing comfort, in that the downside risks associated with a false-positive determination evanesces in the face of unavoidable and impending death. Such considerations regarding false-positive determinations of competence do not indicate that one takes autonomy or the authorizing capacity of moral agents less seriously. It is rather that one recognizes the special fiduciary role of physicians to protect incompetent patients from harming themselves.[2]

These concerns with competency can be brought under two major headings. First, the physician has a concern, indeed an obligation, to act rightly, that is, with authorization from the patient so as not to be blameworthy (e.g., to use the patient without the patient's leave). There is a moral issue at stake for the patient here as well, as Edmund Pincoffs points out in his essay. Namely, the patient is generally a moral agent who must take responsibility for his actions. A determination of incompetency serves to establish when what appears to be an action is in fact a behavior for which the patient cannot be held responsible, placing others in responsibility as fiduciaries. Determinations of competency thus establish whether the patient is responsible for his actions or whether others will have to assume that responsibility by default. Second, the physician has a concern, and indeed an obligation, to act usefully on behalf of the patient. To do this, the patient must be a good judge of his own best interests so as prudently to be able to make choices in terms of his own life plans and projects. Here the focus is not on who is a responsible agent, or on blameworthiness or praiseworthiness, but on consequences. The focus is not on worthiness of happiness but on how to achieve happiness or well-being. This second teleological perspective contrasts with responsibility-oriented deontological concerns.

These concerns with competency support various social practices, which determine when patients are able to choose concerning their own treatment, or when surrogates must choose on their behalf. These practices are central to determine who authorizes the treatment of patients. Out of a recognition of the inability uncontroversially to define the best interests of moral agents, those who understand and appreciate the consequences of their actions, if they do not have overriding obligations to third parties, are usually held to be at liberty to accept and refuse treatment, even if such choices are against what third parties may hold to be the patient's best interests. The social practice of determining competency thus primarily reflects a moral concern that physicians be authorized by patients in order to treat them, rather than patients being reliable authorities as to what constitutes their best interests. Competent refusal of treatment is best understood as the refusal of authorization. However, when patients are not competent, and when they have not provided prior authorizing statements concerning their wishes, and when one does not have evidence of what patients would have authorized, one is left with advancing their best interests as that is generally understood in a particular society. Advance directives prepared by competent persons can thus provide not only authorization or withdrawal of authorization for treatment when such competency is impeded, but they can provide as well a sense of a patient's understanding of his life's projects and his own best interests and therefore of how medical choices are to be understood with regard to that patient. Competent choices by a patient not only provide authorization but a definition of best interests.

This volume looks at these conflicting concerns and interests from different perspectives. The volume is divided into five major sections. The first, 'The Clinical Context', is devoted to the analysis of *informal* competency determinations within the context of the clinic. Though physicians do at times resort to *formal* commitment hearings for the patients they believe to be mentally incompetent, the actual practice of medicine involves numerous informal judgments by physicians regarding patient competency. James A. Knight investigates the heterogeneous character of competency judgments. He argues that competency depends on six conditions: (1) awareness of the nature of one's situation, (2) a factual understanding of the issues with which one must deal, (3) an ability to manipulate information rationally in order to reach a decision on these issues, (4) the ability to function in one's own environment, (5) the ability to perform tasks that one has always been called on to perform, and (6) the character of being consistent in one's mental status. In evaluating a patient's competency, elements of all of these characteristics are used. However, the situation or circumstance in which competency is questioned determines which elements of the test categories are stressed and which are minimized. In short, judgments of competency are contextual. Like Knight, Edmund D. Pellegrino advances a contextual account of competency. He argues that competency turns on the capacity of a person to make a conscious and reasoned choice at a particular time, in a particular context, and about a particular treatment. Competency includes the capacity (1) to comprehend information, (2) to apply information to one's case, (3) to manipulate information, (4) to select an option, and (5) to communicate one's choice. A patient's rationality should be assessed not in terms of the physician's view of what constitutes a rational choice, but in terms of what constitutes a rational process of choosing, which must be judged in light of the internal and logical consistency of the patient's own value system. In short, competency determinations are context-dependent.

The second section, 'Reexamining Concepts of Competence', explores the major philosophical underpinnings of concepts of competency. Here, Tom L. Beauchamp argues that any adequate answer to the question of what is competence in a particular case must supply answers to prior questions about the nature of specific tasks, the abilities required to perform the designated tasks, and the functions of judgments of competence. What Beauchamp sets out to do is to provide a single, basic skeletal meaning of competence. As he argues, 'X is competent to do Y' always means 'X has the ability to perform task Y'. Competency has a gatekeeping function in a wide variety of contexts and establishes the abilities and levels of abilities required in a situation. Any selection of competency tests and any determination of competency thresholds necessarily rests on moral criteria, the more critical of which are autonomy (issues of self-determination, liberty) and beneficence (issues of best interest or welfare). In his essay, Edmund L. Pincoffs analyzes the moral presuppositions underlying determinations of competence. He argues that the duty to inform a patient extends beyond the duty to make the patient fully aware of the possible or probable consequences for himself of accepting or rejecting the prescribed treatment. It also extends to making the patient aware of the facts that bear on the responsibility or irresponsibility of his decision. He concludes that it is the process (and not the conclusion) of reasoning that is of significance. This view involves the recognition that moral agents are responsible for their individual decisions and it is this that matters in judging competent decision-making.

The third section, 'Fashioning Law and Public Policy', investigates the jurisprudential presumptions of the practice of competency determinations. E. Haavi Morreim explores competency as it stands at the interface of law, medicine, and philosophy. She argues that the criteria of competence that a physician applies in clinical practice differ from the standards found in law. The divergences arise from differences between the overall goals of medicine (e.g., eliminate, mitigate or accommodate to dysfunction of mind or body) and those of law (e.g., protecting citizens from improper or unauthorized intrusions into their property, bodies and lives, as well as providing a public forum for the definitive resolution of disputes and the enforcement of solutions). Although medicine and the law share a commitment to holding adults competent until proven otherwise, medicine is better able to handle particular cases of dubious competency because of its commitment to the care and cure of patients and the treatment of their complaints. In either case, as John A. Robertson argues, competency is the watershed line for exercising autonomy across a wide range of areas of decision. On this view, competency points the way to a solution, but it is not the solution itself. The notion of competency is a filter or screen that channels our thinking by limiting the alternatives and factors to be considered. Competency is one point on a rough map of the difficult terrain of life; it signifies the capacity to perform the task of exercising autonomy. Patricia White and Susan Hankin Denise also discuss competency from the perspective of law. They show that there is no single test of competency in the law. Rather, competency is defined in different ways for different purposes. As a result, a person can be legally incompetent in some areas while remaining legally competent in others. For example, the law sets a lower degree of competency for making wills than it does for the execution of agreements such as are framed within the physician-patient relationship. White and Denise do not urge a single view of competency but rather suggest that different interpretations of competency serve different purposes.

The fourth section, 'Case Studies: Seeking Insight from Application', scrutinizes the relationship of conceptual assumptions regarding competency within the informal practices by which physicians decide to treat patients as incompetent to choose, and then proceed to make choices on behalf of patients with or without the guidance of family members. Eugene V. Boisaubin discusses various cases of competency judgments from the perspective of an internist. He views competency as a capacity to act voluntarily with sufficient information. The most useful way to focus a physician's energies in testing for patient competency in a clinical setting is by attempting to determine whether the patient actually understands and appreciates the information necessary to provide an informed consent. If a positive determination can be made in this regard, a patient may be said to have the capacity or competence to decide on a continued or future course of treatment. Marc Perl adds to the reflections of Dr. Boisaubin by discussing competency determinations from the perspective of a psychiatrist. He emphasizes that competency determinations may be more complicated than they appear to the average physician. For physicians in general, strong emotion (e.g., fear, denial, anxiety, sadness, and fluctuations in opinion) are seen to be generally detrimental to competent decision-making. Perl argues that this view is overly simplistic. Competent decision-making is complex and involves more than rational deliberation. Moreover, physicians ought to avoid imposing their own standards of competency on their patients, and this can in part be accomplished by reassessing the role of emotions in such decision-making.

Edwin R. DuBose and Earl E. Shelp argue that standards of competency are continually evolving, in part due to the ways in which we resolve conflicts between expressing concern for and respecting others. DuBose and Shelp urge that the perceived dichotomy between autonomy and beneficence be replaced with an ethos in which the expression of respect for the person *and* the promotion of his care become mutually compatible in the therapeutic relationship. Their hope is to ground a notion of competency that is more than 'minimalist'. This hope is expressed as well by the President's Commission for the Study of Ethical Problems in Medicine and Biomedical and Behavioral Research in *Making Health Care Decisions* [4], when it concludes that although adult patients are entitled to accept or reject health care interventions on the basis of their own personal values, "patient choice is not absolute" ([1], p. 3). Self-determination and patient well-being are seen as fundamental values that should be mutually promoted in medicine.

The fifth and final section, 'Commentary and Critique: Another Look at Concepts of Competence', reflects on the views offered by the previous authors in order to clarify concepts of competence. Harold Y. Vanderpool examines definitions of competency. He indicates how the world-views (belief systems) of patients and their temperaments (emotional states) have critical bearing on *when* they are considered competent, incompetent, or semi-competent. He calls for critical and sustained scrutiny of medical presumptions and traditions regarding how persons are viewed as less than fully competent. To this, Virginia Abernethy offers a cultural perspective on competence. She argues that judgments about competence reduce to judgments about what types of harm society is more willing to risk. She predicts that society will become less willing to risk violating the meaning and spirit of American legal theory which presumes that every adult is competent and should not be touched without his consent. This is the case because of our society's disenchantment with medical technology, beliefs that persons are individually responsible for maintaining health, and the typical perception of resource scarcity.

Stephen Wear worries that many of the analyses in this volume (e.g., by Knight, Pellegrino, Beauchamp, Robertson) support the view that judgments about outcomes and risk-benefit ratios are to be integrated into competency testing. He argues that such heightened monitoring and testing for competency in light of possible adverse outcomes to the patient fundamentally contradicts the basic principles and insights of the ethos of *respect for persons*, and that the strength of this ethos should be sufficient to deny the claims of any rival medical ethos. This is the case because the right to manage one's own affairs with minimal impedance and oversight is fundamental to a secular notion of the moral community. Like Wear, Edmund L. Erde argues that outcome assessments of competency are problematic. Judgments of competence should be made on the basis of the person's performance — evaluating closely associated tasks with reference to the meaning the person's experiences will have for him, rather than by surreptitiously including welfare-oriented consequences, in the meaning of competency. This is the case because individuals are typically the best judges of their own best interests. Moreover, because of autonomy-oriented considerations, it is wrong to impose one's view of best interests on another.

Kenneth F. Schaffner argues for a "sliding scale" model of competency. He forwards three interacting clusters of factors for analysis and consideration with respect to a competency determination: rationality (or understanding), volitional freedom, and aggregated expected benefits and/or harms. This allows one to imagine (and weigh) the interaction of these considerations which on their three axes define an "intellectual space" in terms of which a competency determination is properly made. This image emphasizes that in making a judgment of competency one should weigh each of these factors. Thus, for example, a low risk/high gain procedure may be refused by a patient only if he is located in a region of high understanding and high volitional freedom. Schaffner's point need not be understood as one that in any way undercuts the importance of autonomy, but one which underscores the importance of avoiding false-positive determinations of competence when the costs of a false finding could be high.

In the end, the ways in which one weighs factors in judging competence will depend on the moral and philosophical theory one brings to the task. At root, the moral concern in medicine, as well as in law, is to act with moral authority and achieve the best interests of those in one's care or one's charge. Because we live in a secular society, authority for common action cannot be derived through appeals to God. Nor does moral reasoning possess the capacity to disclose *the* canonical ranking of goods that would need to be accomplished in order to discover a patient's best interests (and thus, for example, to decide between a physician's and a patient's conflicting views of the patient's best interests). One needs to know, for example, how to rank liberty goods and consequences, prosperity goods and consequences, security goods and consequences, etc., in order to determine which choices will maximally protect or achieve the patient's best interests. To make such judgments, one would have had to possess a correct canonical ranking of goods or proper moral sense. But there is no way to establish which moral sense, set of moral intuitions, thin theory of good, view of human nature, or ranking of consequences is morally canonical without at least tacitly endorsing a particular set of assumptions, which is to say, without begging the question. In our secular pluralist world, there does not appear, in short, to be a morally authoritative standard to which one can appeal to resolve disputes about how to judge best interests ([2], [3]).

Moral authority in the postmodern world is least problematically derived from common consent. The authority for joint actions then derives from neither God nor reason, but from common agreement. Hence, there is a moral salience of such practices as free and informed consent, limited democratic procedures, and the free market. So, too, considerations of autonomy trump concerns about best interests because material consent conveys authority even when one cannot discover *the* correct account of best interests. Still, concerns about consequences understood in terms of a patient's likely judgments when competent underscore the need to be clear that a competent decision has been made and moral authority for action has been conveyed.

The essays disclose a tension at the root of ethics, which displays itself in conflicts regarding the nature and significance of competence. One is concerned about the competence of a patient because of two considerations: 1) a patient's choice conveys moral authority; and 2) an individual is usually the best authority regarding his plan of life and reasons for living. It is within a patient's own life, as expressed by the patient, that the consequences of any diagnostic or therapeutic interventions must be assessed, at least in general secular terms. One wants patients to understand and appreciate the consequences of their choices so that they validly provide authority for action. But, one also wants patients to understand and appreciate the consequences of their choices in terms of their stable values so that they function as competent reporters of the context within which diagnostic and therapeutic consequences must be assessed. The first concern, the concern for authorization, focuses on considerations of autonomy. The second concern, the concern to understand a choice in terms of the patient's stable values, focuses on considerations of beneficence. As the essays, and indeed life in general, show, the two will inevitably conflict.

March 2, 1991

NOTES

[1] I am in debt to H. Tristram Engelhardt, Jr., for his numerous suggestions regarding this introduction and this volume.

[2] A similar, but still somewhat different, approach is provided by James Drane [1].

BIBLIOGRAPHY

1. Drane, J.: 1984, 'Competency to Give Informed Consent', *Journal of the American Medical Association* **252** (August 17), 925—927.
2. Engelhardt, H. T., Jr.: 1986, *The Foundations of Bioethics*, Oxford University Press, New York.
3. Engelhardt, H. T., Jr.: 1991, *Bioethics and Secular Humanism*, Trinity Press International, Philadelphia, PA.
4. President's Commission for the Study of Ethical Problems in Medicine and Biomedical and Behavioral Research: 1982, *Making Health Care Decisions: The Ethical and Legal Implications of Informed Consent in the Patient-Practitioner Relationship*, U.S. Government Printing Office, Washington, D.C.

PART I

THE CLINICAL CONTEXT

JAMES A. KNIGHT

JUDGING COMPETENCE: WHEN THE PSYCHIATRIST NEED, OR NEED NOT, BE INVOLVED

Although competency is a legal concept and, under the law, can only be determined by a judge, the clinical realities of patient care require that physicians often make their own assessments of whether a patient is competent or not [19]. 'Psychological competency' rather than 'legal competency' is actually what physicians are determining. Although competency and capacity are used interchangeably, capacity properly is a narrower concept than competency.

One does not have to remind physicians that competent and incompetent functioning can seldom be neatly assessed. As Jay Katz [24] has emphasized, it is a question of more or less, of one and the other, aggravated by internal and external factors. Because the external factors affect the balance between competence and incompetence, and do so to significantly differing degrees, the question must always be posed: "incompetent for what external purposes?" The usual policy adopted by physicians is to presume competence, and only in an occasional and well-specified circumstance seek authorization to override the patient's wishes because of incompetency.

Physicians are generally aware that there are degrees and areas of competence and incompetence. They usually heed the admonition given them in their training that 'competent' and 'incompetent' are not to be used as global attributes of a person but rather be applied more narrowly with respect to particular abilities. They are aware that an individual may be considered competent for some legal purposes and incompetent for others at the same time. For example, a person may not be competent to testify in court but be competent to consent to treatment. Thus, the core meaning of competence is seen as the ability to perform a specific task [13].

It must be acknowledged that judging competence on the basis of the ability to do some specific or explicit task is being challenged. Abernethy has recommended that the presumption of competence should be overcome only when significant dysfunction in an *array* of cognitive and interpersonal domains can be demonstrated. She goes on to emphasize that good overall cognitive skill would be incompatible with a judgment of incompetence, no matter how (by what intrapsychic process) or what the patient decided about medical interventions. A patient who was informable and cognitively capable of making ordinary decisions on matters unrelated to the crisis at hand would be held competent to refuse or accept medical interventions ([1], p. 57).

If the standard recommended by Abernethy were adopted, it would modify the law regarding competence, for the standard of competence is attached to the concept of competency *to do some explicit act*. In establishing the fact of incompetence, evidence must be presented that assesses the ability of the person to carry out the activity in question with sufficient understanding to evaluate accurately such consequences as risks, benefits, obligations, or legal jeopardy [20]. To move away from judging competence as it relates to some specific act may more adequately protect the privacy and independence of the person but possibly weaken or thwart the powerful impulses for social and altruistic concern for others [42].

The nature of medical practice is such that genuine reflection and analysis regarding patient competency are seldom exercised. There is an air of urgency about all treatment, a bias toward action without delay. Any factors that tend to slow down or add extra steps to the process are pushed aside or ignored, unless of real significance. Such an approach in medical practice may be seen in part as convenience for the doctor, but it is more than that. Economic factors loom large and anything that lengthens a costly hospital stay or calls for extra consultations is not welcomed. The physician is tied close to 'the post of use', like the horse is tied snugly to the hitching post. There is little room for movement, for activity that is sometimes seen as splitting hairs.

I. THE GENERAL PHYSICIAN JUDGING COMPETENCE

The physician in the practice of medicine usually approaches the matter of competence or incompetence in an informal manner. Rarely would he be able to delineate a specific set of guidelines for establishing a diagnosis of psychological incompetence. Of course, certain groups of patients immediately alert him to the possibility or probability of incompetence — groups such as the aged, the mentally ill, the retarded, the critically ill or injured, and those facing major surgery. He is alerted to a consideration of competence primarily in his treatment planning and in his overall care or 'management' of the patient. In routine medical treatments and minor diagnostic procedures, consent is assumed and the matter of the capacity to give informed consent is usually not brought into a conscious consideration. Treatment refusal, however, even in minor medical matters would usher in the issue of competence. A general observation is that competence is typically questioned only when a patient refuses to consent to a recommended treatment. All patients are expected to be 'good' patients and fulfill the duties of a patient as emphasized by Talcott Parsons: (a) to maintain a strong desire to get well; and (b) to cooperate with the prescribed treatment regimen ([34], pp. 436—439). When the patient deviates from this expectation, the patient's competence is usually assessed first by the attending physician caring for the patient and then by a psychiatrist who is called in for a consultation.

The patient's refusal to give consent for an operative procedure that would be life-saving or at least life-extending leads the physician almost universally to seek a psychiatric consultation to assess what is 'going on' with the patient. The physician also seeks from family members, if available, their assessment of the patient and the situation. Is the treatment refusal irrational behavior growing out of stress, fear, mental confusion, efforts to control the doctor, depression, hopelessness, or a wish to die? [7] This is the type of question that enters the doctor's mind, and he begins to weigh the different facets of such a question while seeking help from the psychiatrist for a thorough evaluation of the patient.

Furthermore, physicians who are not psychiatrists are often the ones who take the initial steps in the involuntary hospitalization of the mentally ill. This authority is granted to any physician through state mental health laws and can be exercised through the use of the so-called 'physician's emergency certificate'. Usually the patient suspected of mental illness is brought to the physician by another person. The physician determines if the patient is suicidal, homicidal, gravely disabled mentally, or unable or unwilling to seek voluntary admission to a hospital. If one or more of these conditions exist, then the physician fills in a printed form known as the 'physician's emergency certificate', and has the patient admitted to the hospital. The physician, in the assessment of the patient, takes a history and does a mental status examination with emphasis on orientation, mood, thought content, and especially the presence of any delusions or hallucinations. The mental status examination is an examination of the functioning of the psyche and classifies and describes all areas and components of mental functioning that are involved in the diagnostic classification of mental problems. In such situations the physician's assessment of competence is somewhat more formal than in other situations because the mental status examination is a required part of such an assessment. Furthermore, his attention is drawn sharply to an evaluation of both the behavior and thinking of the patient.

Another situation calling for concern is where the patient is in need of major treatment, such as a surgical operation, and willingly gives consent, but at the same time the physician suspects incompetence. Among the conditions in which incompetence could appear are brain tumors, the aged in unfamiliar circumstances, metabolic derangements or chemical imbalances of the blood, limited intellectual resources, and severe stresses. If the incompetence is not too obvious, doctors are often accused of considering the patient's consent as informed. If the incompetence is obvious to most anyone who sees the patient, or reads the patient's record, the physician, 'to protect' himself and his patient, would get a psychiatric consultation and then seek judicial consent. Possibly this suggests that the physician's concern with competence is only present when the physician is forced to consider it because of internal or external factors. Most likely, this is not a fair assessment of the physician's concern with competency of patients.

As for the physician who is not a psychiatrist, he usually does not assess competence unless treatment refusal or lack of cooperation brings the issue to the forefront or the patient's behavior or comprehension is obviously suspect. At this point probably a brief mental status examination would be done. After the physician decides the patient is or is not competent, he may then obtain a psychiatric evaluation to check his clinical judgment in the matter and to decide on further steps in patient management.

II. THE PSYCHIATRIST AS CONSULTANT IN JUDGING COMPETENCE

The psychiatrist is readily available in most hospitals for consultation with other health professionals and is frequently consulted in matters of patient competence. In some ways, the psychiatrist's decision is often just as important as that of decisions emanating from the bench regarding competency [19]. The psychiatrist is frequently called on to decide if a judicial determination of competency is warranted. This is so because an immediate resort to the courts whenever the question of incompetency arises would be too expensive and time-consuming. Furthermore, once the case reaches the court, the psychiatrist's assessment often serves as the major source of data for the judge's decision [33].

Medical and surgical colleagues in requesting a consultation from a psychiatrist, when a patient's competency is in doubt, may actually be seeking from the psychiatrist 'permission' to proceed with the treatment without involving the courts [5]. If this is true, does this expectation exert a special influence on the psychiatrist that would bias his decision about whether to declare the patient competent or incompetent? Many of us in psychiatry, in doing consultation work, have felt this influence and may often build our argument to support the assessment of competence or incompetence in part, at least, in response to this expectation. At other times, when faced with a patient's refusal of a major operation, the psychiatrist's colleagues ask: "Don't you agree with us that this patient is psychologically incompetent?"

A difficulty with the psychiatrist in interacting with the legal system is the danger that he will abandon the uncertainties and ambiguities of the clinical perspective for the seductive rationality of legal thought. As legal scholars remind us, the presumptions of the law are abstractions that, although useful in the system where they originated, are inexact approximations of the reality that the psychiatrist daily experiences. Thus, not infrequently the psychiatrist may be led incorrectly to assess a patient's functioning at a single time, in a single setting, with an uncertain factual base, and then draw a global conclusion about a patient's functioning. Often the psychiatrist believes that the law tends to address competency as a fixed attribute of a person. The psychiatrist may not always be aware that what the law calls competency is, in fact, a set of deductions from a variety of clinical data that can be as subject to influence and change as the more basic mental attributes on which it is based [5]. In actuality, law or the courts do consider all of the patient's characteristics in making what in law must be a *yes* or *no* decision at a particular point in time.

III. DISAGREEMENTS AMONG PHYSICIANS IN JUDGING COMPETENCE

Do physicians disagree with one another about the competence of a patient? When incompetency is fairly obvious, there is seldom disagreement. The greatest disagreement comes in situations where a patient refuses a well-established treatment method for a life-threatening condition — a condition that is likely to hasten the person's death or cause serious disability. The disagreement may encompass ethical, legal, and medical dimensions. The following case from the surgical service of a large general hospital illustrates the complexities of assessing competence, the basis for disagreement, and the value systems of both the patient and the doctors.

A 55-year-old patient came to the hospital because of a growth in the back of his mouth. He was found to have a malignant tumor of the soft palate and tonsillar area of the mouth and throat. Physicians in the specialty of otolaryngology informed him of the nature of the tumor and their recommendation of a surgical procedure involving a radical neck dissection, followed probably by radiation therapy. The patient was told that his condition was serious but not without hope. Possibly the tumor could be brought under complete control, but if that were not accomplished, then, at least, the time he had remaining would be made more comfortable by the prescribed treatment. If he received no treatment, the physicians predicted continued growth of the tumor, and, in time, marked interference with the patient's respiration, food and fluid intake, swallowing, and speech, resulting in an early death accompanied by enormous suffering.

The patient listened attentively to the treatment plan, was given an opportunity to ask questions, and was checked to see if he understood what would be done and the expected consequences, both with and without surgical intervention. The patient refused the treatment plan in totality and said that he understood what was recommended but that he preferred to take his chances and receive no treatment. The picture of a grim and dismal prognosis did not change the patient's mind.

The otolaryngology staff requested that the consulting psychiatrist evaluate the patient and the situation. They felt that the patient had to be emotionally disturbed, and, in fact, implied that the patient had to be 'crazy' to refuse treatment when he was clearly informed of the consequences.

In the evaluation of the patient and the situation, a critical question related to the patient's psychological competence. A careful examination showed that the patient's decision-making competency was intact. He was oriented in all spheres, had good recent and remote recall, and evidenced no confusion in any area of his thought or life. The quality of the patient-professional relationship was explored, and the patient gave all indications of being on excellent terms with doctors and nurses. Also, the patient expressed no dissatisfaction with the hospital and its overall care of him by doctors and nurses. Although the patient was depressed, the depression was not deep enough to interfere seriously with his decision-making ability.

It was difficult for the physicians in otolaryngology to accept the psychiatrist's evaluation that the patient was psychologically competent. They felt that the patient's decision to refuse treatment was an unreasonable, irrational act. The life-and-death situation, they contended, imposed enough stress on him to interfere profoundly with the requisite mental and emotional stability to make an informed choice. They went on to say that in refusing treatment and leaving the hospital, the patient would be back soon with far advanced cancer, and they would be the ones to take care of the patient with massive feeding and breathing problems in the final stage of his life.

Discussion centered on the patient's autonomy and his right to refuse treatment if competent, a right grounded in sound ethical principles, as well as those of constitutional and common law. Immanuel Kant, for example, argued that a rational person should be autonomously self-regulating. He expressed his view of the centrality of autonomy when he stated that freedom is autonomy, meaning "the property of the will to be a law in itself" ([23], p. 65).

Of course, the loophole in the right to refuse treatment is that the patient must be competent. Questions may surface about competency, and disagreements may arise. In this case, the surgeons doubted the patient's competence and found it difficult to accept the psychiatrist's assessment that the patient was psychologically competent. The courts wrestle with the same issue. For example, in 1966, in a case in Washington, D.C., the court put forth a competency test:

'Can the patient understand the situation, the risks, and the alternatives? It does not matter whether the patient's choice is rational or not, as long as the patient is able to make a decision' ([8], pp. 180—182).

Of course, such a court decision raises questions about the nature of competency and rationality. An important point, however, is that, in ascertaining whether or not a patient is competent, one must remember that it does not matter whether the patient chooses as *we* would choose in a particular situation, or whether the patient would choose as we would like him to choose. Whether we consider a particular patient's choice to be a rational one in the circumstances is not the issue, but it still philosophically may be requisite that the patient's decision from the patient's own framework be rational in order to be competent.

To take the matter of rationality a step further, one should remember that 'rational' and 'irrational' are terms used as attributes of the decisions made by patients [12]. An irrational decision is one based on an irrational *belief*, such as the patient's not believing he has cancer in spite of overwhelming evidence that he has, or on an irrational *desire*, such as a patient's preferring to die rather than have a gangrenous foot amputated when he can give no reason for that choice. Irrational desires may include the desire to die, to suffer pain, to be disabled, or to be deprived of freedom. The crux of the matter is that it is not always irrational to desire these things, but it is irrational to desire them *without* some adequate reason ([13], [18]).

Further discussion about competency is indicated at this point. In cases such as this one, questions of competency arise, in a general or specific sense. Ethical or legal issues impinge on each determination. Some guidelines include four areas that need exploration: (a) Does the patient understand the condition for which the treatment is proposed? (b) Does the patient understand the nature and purpose of the treatment? (c) Does the patient understand the risks and benefits involved in undergoing the treatment, including the effects? (d) Does the patient understand the consequences (risks and benefits) of alternative forms of treatment or of *not* undergoing any treatment at all? [10] These areas of exploration are similar to those included in the doctrine of informed consent. Other guidelines are related, but often contain some additions. For example, Gutheil and Appelbaum urge a careful examination of the patient's use of rational processes for arriving at a decision. This could include the basic components of the mental status examination — orientation, memory, intellectual functioning, judgment, impairments of rationality (delusional thinking, hallucinations), and mood alterations [19]. Further, they emphasize that spurious indicators of possible incompetency must not be overlooked such as the psychodynamic factors involving deep fears that invoke emotional defenses, for example, when a patient faces a serious operation involving cancer or when the situation includes factors such as conflicts and misunderstandings between the patient and doctor. A patient such as this one was unable to control his malignancy and maintain a sense of autonomy and self-determination in his life. Although he could not control his malignancy, he could control the doctors. Thus, was the treatment refusal used by the patient to bring him an increased sense of autonomy and control?

No stone was left unturned in evaluating this particular patient. At the same time, while it was the psychiatrist's clinical judgment that the patient was competent, not all the members of the health care team, especially the surgeons, shared that judgment. Of course, there is a strong societal and health professional bias in favor of treating treatable patients, and such biases can easily influence one's clinical judgment about competence. Treatment refusal in patients such as this one can easily become tantamount to incompetency.

What about the views of family members regarding a patient's decision? This particular patient had no immediate family, and he felt that he was a free agent in deciding his course of action, because nobody was dependent on him for anything, emotionally or economically. In fact, the great libertarian, John Stuart Mill, believed that coercion was justified only on the grounds of preventing harm to others: "[The person's] own good, either physical or moral, is not a sufficient warrant." It must not be forgotten, however, that Mill went on to state that "it is, perhaps, hardly necessary to say that this doctrine is meant to apply only to human beings in the maturity of their faculties" ([31], p. 73).

While this patient's refusal to accept possibly life-saving therapy was not interfered with, many on the hospital staff felt that this should have been otherwise. Their arguments centered on a variety of points. Could the patient be depressed, ambivalent about the proposed treatment regimen, or lack confidence in the hospital and its staff? Some regarded a refusal of treatment that could possibly save his life or at least extend it as clearly unreasonable, even if competently decided. Further, they argued that the state has a 'compelling interest' in preventing this patient's death. Others took a strong paternalistic position: the competent adult person's liberty of choice should be limited for his or her own good in order to prevent harm from befalling the person [28]. The argument that most of the patient's physicians put forth challenged the patient's refusal on the grounds that a patient in a life-or-death situation, by that very fact does not possess the requisite mental or emotional stability to make an informed choice. To put the issue another way, they were taking the view that the patient's autonomy was impaired by his illness and his reactions to the illness. As T. F.

Ackerman would express it, the issue in a more general sense relates to the components in a patient's illness that can involve serious constraints to his autonomous behavior [2].

Needing examination and qualification is a statement often heard from physicians that a patient in a life-or-death situation does not possess the requisite mental or emotional stability to make an informed choice. Such a global statement could not be defended in a careful study of individual cases. Each person reacts differently to information or 'personal bad news' because of a host of factors including one's personality profile, life's circumstances, and intellectual or emotional resources. With proper or fortunate interventions, the expected outcome from one very serious condition may be quite different from that of another. To say unequivocally that severe stress automatically renders a patient's decision invalid or incompetent may be a cop-out opposed to trying to understand and accept a decision that seems contrary to one's own view of what is reasonable and appropriate. Actually, a careful evaluation of patients in life-and-death situations reveals a range of responses from a high level of competence to disorganization of personality. Thus, each case must be evaluated individually, for general maxims often do not fit a particular case. In other words, it behooves attending physicians to assess each individual case on its own merits and not be guided too readily by a principle that on the surface does deserve some attention. With the possibility of denial in mind, the physician is obligated to screen carefully for both general and specific competency and weigh the magnitude of denial in the patient's decision-making process.

The question remains: why did this patient refuse treatment? What reasons did he give for refusing treatment? He was not willing or able to discuss in detail his refusal. He saw the operation as mutilation, as causing considerable disfigurement of his face and neck. He abhorred the thought of that. At some level, he seemed to think that the doctors were wrong in their diagnosis and pessimistic prognosis and that 'the thing' that was happening to him would pass away on its own accord. He never articulated this openly but implied it. Thus, one has to raise the question of how denial, fueled by hope, was operating in this patient, and how much of a factor it was in the patient's refusal of treatment. At the same time, the psychiatric examination revealed a patient who appeared to be in control of his thinking and not strongly influenced by factors beyond his conscious awareness.

Denial is not always easy to assess in a patient such as this one because it could range from his choosing not to deal with an unpleasant circumstance to his automatic and involuntary exclusion from awareness of a disturbing aspect of reality, as well as an inability to acknowledge its true significance. The physician has to make a serious effort to determine the level of the patient's awareness of his condition and what the future holds. The patient's choice may be questioned more seriously if an unconscious determinant is operating rather than one close to the surface of conscious reality. Clinically, the physician must assess denial and weigh carefully the hidden or surface dimensions of the patient's refusal behavior. After a careful appraisal, what was thought initially to be denial may be reality-based and the result of a choice that is conscious and fully aware. Outright denial of illness will be discussed later.

A difficult and mutilating surgical procedure was recommended in an effort to save this patient's life; and a favorable result was far from being a certainty. In other words, the risk was great and the benefit was in question. One can ask how much the risk/benefit ratio influenced the decision to honor the patient's refusal of treatment and its consequences. If the risk had been minimal and the benefit great, such as offering a cure for his cancer, probably the patient's competence to refuse treatment would not have been as easily honored.

The way medicine is generally practiced in the United States, the risk/benefit ratio is a major factor in determining the competency of a particular patient in accepting or refusing treatment. If, for a serious illness, the risk in treatment is minimal and benefit is great, the patient who refuses treatment is likely to be declared incompetent. If risk is great and benefit is minimal, the patient who refuses treatment is likely to be declared competent. Of course, the patient's motives and the time (or lack thereof) to assess them play an important role in assessing competency. This is especially so in emergencies or in situations where immediate treatment seems indicated, and there is little or no time to assess factors that may be fueling a refusal of treatment or ambivalence about it.

For example, an often cited clinical case is the patient who comes to the emergency clinic complaining of severe headaches of recent origin and is found on examination to have acute pneumococcal meningitis [11]. The patient refuses the proposed treatment of penicillin plus hospitalization. Already the patient is developing some clouding of consciousness. In such a case where the lack of treatment would likely be fatal and treatment almost certain to be successful and without sequelae, the physicians will waste no time in having the patient declared gravely disabled or incompetent and initiating immediate treatment. As already mentioned, time will not permit an assessment of motives for refusing the treatment. Further, the well-recognized nature of the illness is already producing its damaging effects on the brain, compromising the patient's decision-making capacity.

The right of autonomous decision-making on the patient's part may come in direct conflict with a right claimed on the physician's part: the professional duty and right of the physician to do what correct medical practice dictates. In the practice of medicine, it is taken for granted that the physician has a right not to have restraints applied to a therapeutic program that the physician regards as necessary for the patient's welfare or survival [11]. The physician, like society in general, sees human life and the duty to preserve it as paramount moral values. Thus, the physician is biased toward compelling medical treatment because of these values and will quickly invoke them when treatment intervention is challenged. Of course, other values for a particular patient may take precedence over those of the physician and society in general. For example, the patient who is a Jehovah's Witness may be risking the loss of his life by refusing a blood transfusion but in doing so he is not jeopardizing his chance for immortal life with God and the faithful. In other words, the Jehovah's Witness is weighing eternal life over against mortal life, and because of his strong metaphysical belief, he may opt to allow his mortal life to terminate. Since the person who holds this religious belief may be competent and rational, there is no ground for paternalistic intervention by the physician, when paternalism is characterized by Gerald Dworkin as: "the interference with a person's liberty of action justified by reasons referring exclusively to the welfare, good, happiness, needs, interests, or values of the person being coerced" ([15], pp. 64—65).

Ruth Macklin, in arguing for respecting the autonomy of patients as rational persons, emphasizes that autonomy implies a responsibility for one's decision — a responsibility that entails acceptance of the consequences [26]. Thus, these consequences include the right of physicians to reject a treatment regimen proposed by the patient if that treatment regimen is contrary to sound medical practice. If faced with the consequences, the patient may opt for treatment. Those patients who remain steadfast in their refusal are exercising their right of autonomous decision-making in matters concerning their own welfare. Risks to health or life that a person may take in the interest of something he considers worth the risk appear to know no bounds. If the person is rational and competent to make decisions, the risks are his to take [26]. In such a situation, the physician knows the patient's motives and has had time to assess them if they were not otherwise clear [11].

It is obvious that illness can interfere with autonomy. As Cassell states, the sick person is not just a well person with an appended illness. The sick person is different from the well to a degree dependent on the person, the disease, and the circumstances in which the person is sick and is treated [11]. One can safely say that illness interferes with autonomy to a degree dependent on the nature and severity of the illness, the person involved, and the setting. Thus, one must weigh carefully what autonomy means to a sick person, when autonomy requires both authenticity and independence, as Dworkin argues [14]. Authenticity is the true selfness of a person, wherein a person's beliefs, ideas, or actions are unique despite their source. Independence involves, above all, freedom of choice. Freedom of choice involves knowledge about the area where choice is to be made, the ability to reason, and the ability to act on one's choice. In reality, knowledge is lacking or seldom complete for the sick person, and reason is often impaired, especially in the face of pain, or suffering, and an uncertain future. Further, illness interferes profoundly with the ability to act, whether the patient is bedridden or not. Thus, the stage is set for the doctor to act on behalf of the patient, when the wishes and actions of the patient do not appear to be truly autonomous.

IV. THE FAMILY'S ROLE IN JUDGING COMPETENCE

Physicians see the role of the family as exceedingly important in decision-making regarding the incompetent or questionably competent patient. Often physicians will seek advice from family members if uncertain of the patient's capacity, in an effort to utilize family assessments or viewpoints (sometimes described as 'alter wills') to help in assessing the person's autonomy [22]. Family members are given special rights and responsibilities in the care of each other, and strong arguments can be made for the importance of the family's role as agent for the patient when the patient is incapable of making decisions. The question is often raised whether this can be justified, morally or legally, without a court determination of incompetency. Nevertheless, by custom if not by legal authority, physicians and the courts both turn to members of the patient's family ([41], pp. 43—48).

It is generally accepted that the decision made for the *no longer* competent patient should approximate as closely as possible the decision that the patient would have made had he been able to decide for himself. Since the decision should recognize or incorporate the value and belief system of the patient, the family members are normally in the most reliable position to know those values and to act on them. The ties that bind family members are ones of responsibility, including a mutual responsibility to protect one another's welfare. In those cases where disagreement is strong among family members and the decision is critical, elevating the decision to the court is the most prudent course to follow. In actual practice, however, even in situations where family members agree, hospitals and staff physicians may not be willing to go along with the wishes of the family without a court order. For example, a young adult woman became acutely psychotic and was diagnosed as having a brain tumor. Although incompetent to give consent for an operative procedure, she was quite willing to have the tumor removed, as well as was every member of her family. The neurosurgeon sought a court order before proceeding with the removal of the tumor. The question was raised, however, in a discussion of this type of patient whether the neurosurgeon would have bothered to get a court order if the patient's incompetence to give informed consent had been less obvious and not so much in the public domain.

V. THE PSYCHIATRIST'S APPROACH IN JUDGING COMPETENCE

Psychiatrists have had a long history of involvement in competency evaluations, and recently systematic efforts have been made to study the issues involved ([4], [5], [6], [19], [37], [38], [39]). Presentday concerns with the issues surrounding the rights and responsibilities of patients regarding treatment and treatment refusal have led to a more intensive consideration of the dimensions of both general and specific competency.

Approaches to the Evaluation of Competency Regarding Treatment

Many approaches have been used to determine competence. The

President's Commission for the Study of Ethical Problems in Medicine and Biomedical and Behavioral Research designated three: (1) outcome, (2) status, and (3) function [36]. In the 'outcome approach', decisions that do not reflect community values are used as evidence of incompetence. In the 'status approach', a person's competence is based on the physical or mental status of that person (consciousness, age, mental or physical diagnosis). The focus in the 'functioning approach' is on the person's actual functioning in decision-making situations. Most commentators, as Annas and Densberger have emphasized, have assumed that the 'functioning approach' is the proper one, and have tried to define in the best possible way the attributes needed to function competently [3]. Further, Annas and Densberger state that the 'functioning approach' is the most reasonable and appropriate in hospital and nursing home settings, and it has a solid legal pedigree in the context of treatment refusals.

Appelbaum, Roth and their associates, in using the 'functioning approach' to evaluate competency, show that the various efforts to define the standards for competency generally cluster into four groups ([5], [6], [39]). These four groups are: (1) evidencing a choice, i.e., communicating a decision about one's participation, (2) factual understanding of the issues, including the cognitive capacity to consider the relevant issues, (3) rational manipulation of the relevant information, i.e., how the information is used by the person in the decision-making process, and (4) appreciation of the nature of the situation, affectively as well as cognitively. (Appreciation goes beyond factual understanding in that it requires the person to consider the relevance to one's immediate situation of those facts understood previously in the abstract.)

One can draw from these clusters a variety of tests for competency. The authors mention five, with the addition of two variables, upon which determinations of competency are often made.

(1) *Evidencing a choice.* This verifies the presence or absence of a decision by a patient for or against treatment. The preference may be yes or no, or the desire that the physician make the decision for the patient. Thus, the patient who does not formulate and express a preference as to treatment is incompetent.

(2) *Reasonable outcome of choice.* This test entails the assessment of the patient's capacity to reach the 'reasonable', the 'right', or the 'responsible' decision. The emphasis is on outcome, and thus the patient who fails to make a decision that is roughly congruent with the decision that a 'reasonable' person in like circumstances would make is viewed

as incompetent. As Roth, Appelbaum, and associates point out, the benefits and costs of this test are that societal goals and individual health are promoted but at considerable expense to personal autonomy. Because the test rests on the congruence between the patient's decision and that of a reasonable person (usually, the physician), it is biased in favor of decisions to accept treatment. The end result in this decision-making process is that if patients do not decide the 'wrong' way, probably the issue of competency will not arise.

(3) *Choice based on 'rational' reasons.* This test category attempts to ascertain the quality of a patient's thinking and whether or not it is a product of mental illness. The psychiatrist uses parts of the mental status examination in this assessment. A danger of this test is not unlike that of the reasonable outcome test (the above category). If the patient decides in favor of treatment, the issue of the patient's competency may not arise because of the doctor's bias toward consent to treatment and against refusal of treatment. Although this test is clinically much in use, it poses some conceptual problems [17]. The problem areas include the difficulty of distinguishing between rational and irrational reasons and of drawing inferences of causation between any irrationality believed present and the patient's decision regarding treatment. A patient's decision could be the same even when cognitive processes were less impaired. For example, a delusional patient may refuse electroshock therapy not because of being delusional but because of being afraid of it, which is considered a normal response.

(4) *The ability to understand.* This test category relates to the ability of the patient to understand the risks, benefits, and alternatives to treatment, including no treatment. The patient's ability to understand may be tested by asking a series of questions concerning risks, benefits, and alternatives to treatment. The critical point in this test is that the patient is able to comprehend the elements that are presumed by law to be a part of the treatment decision-making or informed consent doctrine. How the patient weighs, values, or puts together these elements to make a treatment decision is not important. This test appears to be the most consistent with the law of informed consent [30].

(5) *Actual understanding.* This test category makes competence turn on the accuracy of the patient's perceptions. It requires physicians to make reasonable efforts to ascertain that their patients understand what they are told and encourage active participation in treatment selection. In fact, it tests those issues central to patient decision-making about

their treatment. It requires patients to consider the relevance to their immediate situation of those facts they have previously been given and understood.

In evaluating a patient for competency, elements of all of the test categories are used; however, the situation or circumstances in which competency is questioned determine which elements of the test categories are stressed and which are minimized. In theory, competency is an independent variable, but in practice it appears to be dependent on two other variables: (1) the risk/benefit ratio of treatment, and (2) whether the patient consents to or refuses treatment [39]. Where there is a favorable risk/benefit ratio to the proposed treatment and where the patient consents to the treatment, no reason seems to stand in the way of administering treatment. In such situations, a test employing a low threshold of competence may be applied to find even a marginal patient competent to consent to treatment. On the other hand, when the risk/benefit ratio is favorable, and the patient refuses to consent to treatment, a test employing a higher threshold of competency may be applied. Thus, the patient that at most may be only marginally incompetent is found incompetent, and then consent is sought from a substitute decision-maker to override the patient's refusal of treatment. As is frequently emphasized, when the risk/benefit ratio is favorable, the bias of both health professionals and judges is usually skewed toward providing treatment. Thus, a test of competency is applied that will permit the treatment to be administered irrespective of the patient's actual or potential understanding [39]. At the same time, there is a growing reluctance to permit patients to undergo treatments that are very risky or for which the benefits are highly speculative. Thus, when the risk/benefit ratio is unfavorable, tests may be chosen to help the patient appear competent so that his refusal will be honored or to appear incompetent if he consents to treatment, thereby preventing him from undergoing treatment.

It is obvious that none of these test categories by selective application could be used to determine that a grossly impaired patient is competent or that a normally functioning patient is incompetent. A test of competency, however, can be used within limits to achieve the desired medical or social end when the patient's competency is not absolutely clear-cut. This occurs as a consequence of the strong societal bias in favor of treating patients as long as it does not expose them to serious risks. Judgments about competency are made more complex by the fact that they reflect social considerations and societal biases as much as they reflect matters of law and medicine [39]. At the same time, probably the most precise concept of competence that can be developed is one that encompasses the capacity to understand and appreciate the nature and consequences of one's acts — one's decision to accept or refuse treatment [3].

It is of interest to note that philosopher Bruce Miller, in his focus on the functioning approach to evaluating competence, has listed four increasingly strict standards. His standards are analogous in some respects to those of psychiatrists Appelbaum and Roth: (1) free action, involving a voluntary and intentional choice, (2) authentic decision — a decision that reflects the person's values, (3) effective deliberation — an evaluation of the specific alternatives and their consequences, and (4) moral reflection — reflection on and acceptance of the moral values upon which the decision is based [32]. Of course, there are pitfalls in assessing the individual's personality implicit in the authentic decision and moral reflection tests.

Other Factors in the Evaluation of Competency.

First, one is alerted to work with a low threshold of suspicion of incompetency when evaluating particular populations whose members are at increased risk of manifesting incompetency [19]. Groups in these particular populations include:
— Acutely psychotic patients who often suffer from a clouding of consciousness.
— Chronically institutionalized patients who may have lost the capacity to evaluate proposed interventions.
— Patients with organic brain impairment — that is — a disturbance in the physiologic functioning of brain tissue at any level of organization, whether it be structural, hormonal, biochemical, or electrical — often the geriatric age group, who may move easily in and out of confusional states. Also included here are patients with infections of the brain or meninges, acute or chronic, such as meningitis or brain abscess.
— Depressed patients, even the nonpsychotic ones, whose hopeless, helpless thinking may handicap reasoning about treatment alternatives and plans.

— Mentally retarded patients, whose disability may be increased by a concomitant psychiatric illness.
— Patients who are facing especially risky medical or surgical procedures with the hope of little direct benefit from them.

Screening for general competency is the next step, regardless of whether the concern is with the patient's general competency or one's competency to deal with a specific issue. It is possible for a generally incompetent patient to be capable of deciding about a specific task and conversely for a generally competent person to be specifically incompetent in certain areas. There are distinct qualities usually expected to be present before a patient can be judged to be acting competently [19]:

(1) An awareness of the nature of one's situation. The person should have an adequate sense of the current status of the major elements of his or her life: sources of support, circumstances of living, extent of resources, significant supportive relationships, and any limitations on natural functions.

(2) A factual understanding of the issue(s) with which one must deal. Since a person has to make decisions about the major elements of his or her life, those decisions should be grounded in a clear understanding of the facts involved.

(3) The ability of the person to manipulate information rationally in order to reach a decision on these issues. This assessment includes the basic components of the mental status examination and represents the area where psychiatrists have primarily directed their efforts in assessing competence. (The mental status examination for the psychiatrist is comparable to the physical examination for the general physician.) The important areas of the mental status examination upon which the psychiatrist focuses are: orientation, memory, intellectual functioning, judgment, impairments in rationality (delusional thinking and hallucinations), and mood alterations ([27], [40]).

(4) The ability to function in one's own environment. Competency is not a fixed attribute that remains constant in the face of changing environmental factors. A patient, functioning well at home, may become confused in the new environment of the hospital, especially at night when familiar landmarks are not seen.

(5) The ability to perform those tasks that one has always been called on to perform. If called on to perform tasks one has never performed and fails to do so, the person should not be deemed incompetent.

(6) Observing consistency in the patient's mental status. Because of the rapidly changing nature of many mental states or the conditions that lead to incompetency, if possible a patient should be observed and interviewed on more than one occasion.

An assessment of specific competency usually follows the general competency examination. For example, in trying to evaluate the person's competency to consent to psychiatric treatment, one focuses sharply on the specific aspects of the person's current situation, the proposed interventions and alternatives on his behalf, and how rationally he assesses the information given him.

Spurious indicators of specific incompetency deserve serious consideration in any assessment ([19], [21]). Clinical factors strongly influence the dynamic and labile nature of competency. Psychodynamic factors such as a patient's fear of surgery could evoke nonrational responses. An example is the woman who is faced with surgery for cancer of the uterus and who is afraid that after the surgery she will be abandoned by her husband because she is no longer a 'complete' woman. She may handle the situation by denying or distorting its reality, or even becoming psychotically depressed. Psychopathological factors are similar to psychodynamic ones but differ in that the cause of the fluctuations in the patient's mental status are not always discoverable. Many patients have marked fluctuations in their level of mental functioning, dependent on time of day, medications consumed, concurrent physical illnesses, and the like. Poor compliance by the patient in taking the medicines exactly as prescribed, often unknown to the attending physician, may lead to impaired thought processes. Situational factors such as the following ones must also be considered. Ill-will between the patient and the physician may lead the patient to 'act crazy' to frustrate the physician. The patient's rapport with the examining physician, the patient's perception of the purpose of the examination and what is expected of him, and the nature of the setting in which the examination is carried out are all significant factors that may influence how a patient acts or appears.

The Problem of Denial in Assessing Competency

Patients who deny the presence of illness and who consequently refuse treatment on the grounds that no ailment exists to treat furnish a genuine challenge to the psychiatrist in evaluating competency [38]. The assessment of such competency requires a consideration of the accuracy of patients' appreciation of the nature of their situation. Denial may range from the inability to accept at this time a reality that is intolerable to a psychotic denial associated with some of the major diagnostic categories of mental illness such as severe depression, schizophrenia, mania, paranoid states, and organic brain impairment. Disordered thinking and behavior may be obvious to family and others but not to the patients and furnish no problem in assessing competence. The problem arises, however, for those patients whose delusional thinking is concentrated in one particular area and who present few other signs or symptoms of mental impairment. (For example, a thirty-year-old man who is convinced that he is five months pregnant and who shows no other clinical evidence of delusional thinking or mental disability.) It may be quite unclear whether these patients should be categorized as either competent or incompetent. These people often handle the affairs of their daily lives quite well and are able to assess realistically the risk/benefit ratio of the proposed treatment. The problem is that they do not see that the treatment has any relevance to them because they do not see themselves as ill.

Culver and associates have proposed an approach for assessing a patient's refusal of treatment that may help in the dilemma brought about by a patient's denial of dysfunction [12]. They recognize that a person may competently exercise choice although denying certain aspects of a situation. In an effort to protect autonomy to the extent that it deserves, they argue for a strict cognitive test of competency that would encompass denial. They contend that a patient is competent to decide about having a particular treatment when the following are satisfied:

(1) The patient knows that the physician believes the patient is ill and in need of treatment, although the patient may not agree.

(2) The patient knows that the physician believes that this particular treatment may help the patient's illness.

(3) The patient knows that he is being called upon to make a decision regarding this treatment.

Culver and associates stress that the three requirements for competency listed above are minimal criteria because the intent is to consider a patient competent to make a treatment decision unless he is clearly incompetent to do so. Such a position is in keeping with the insistence of Culver and his associates that 'competent' and 'incompetent' not be used as global attributes of a person but rather be applied more narrowly with respect to particular abilities, for example, that a person be described as 'competent to do X' when certain criteria are met. Thus, a person is competent to decide about 'treatment X' if he meets the criteria listed above.

The three-pronged approach of Culver and associates has much to commend it, but it does not include a vital requirement in the assessment of competency to consent to or to refuse treatment: the accuracy of the patient's appreciation of the nature of his or her situation — both a cognitive and affective understanding. Without the ability to make an accurate appraisal of the presence or absence of illness, the patient could not weigh the benefits of treatment against the risks. At the same time, one cannot deny that serious problems or difficulties are raised by requiring a patient to recognize his illness before deeming him competent to refuse treatment [38]:

(a) The standard hovers close to the 'Catch-22' situation in which a patient's denial of the need for treatment is taken as evidence of that very need.

(b) It represents an intrusion on the autonomous functioning of the patient, such that the patient is no longer able to define for himself what state of existence he is willing to accept as constituting health.

(c) If carried to an extreme, such a standard would enable society to define a variety of nonconforming behaviors as illnesses and then to relieve the individual of the right to decide whether these conditions should be treated.

VI. VALUE ISSUES THAT REMAIN TROUBLING

Ethical and value issues weigh heavily on any clinician in decision-making related to patient competency. There is always the task of respecting and protecting the patient's right to self-determination, as well as promoting rational decision-making, while at the same time exercising the moral responsibilities of the physician to intervene on behalf of patients with reduced autonomy.

The paternalism of the doctor also is not without its hazards. What are the forces that should operate to balance the power between the patient and the doctor? The doctor's responsibility and accountability, especially for the incompetent patient, take on extra dimensions. There is a risk in labeling too quickly the patient as incompetent because it is easier to think one is treating an incompetent patient, despite his protestations, than to think one is forcing his own treatment decision on a competent patient who disagrees with it [12]. At the same time, the doctor who withholds treatment from an incompetent patient who refuses it may be held liable for that patient if the doctor does not take reasonable steps to obtain some other legally valid authorization for treatment. This introduces, in fact, a central ethical problem: who should have decision-making authority in cases where the patient is incapable of deciding for himself or the patient's competency is questioned? The family has an important role here as substitute decision-makers or agents for the patient, but questions can be raised about the transfer of authority to the family and the possible lack of objectivity in the family.

It is generally agreed that a decision made for a *no longer* competent patient should approximate as closely as possible the decision that a patient would have made had he been able to decide for himself. Such a goal is not easy to attain, especially when one would try to incorporate into the decision the patient's value and belief system. Yet, the family is in the best position to assure this. At the same time, the doctor must be on guard not to let the family represent the patient's interest except in those cases in which it is abundantly clear that the patient is unable to represent his own interests.

The doctor is aware that many ask why the question of competency is usually raised only for those patients who refuse medication or other treatments. This has led some members of the bar, working in the mental health field, to see the evaluation of the question of competency as a means of overriding the will of the patient, rather than an effort to ensure meaningful decision-making. This, of course, need not and should not be. As many have stressed, the legal profession, with mental patients, is often reluctant to recognize the legitimate role that paternalism must play in a caring society [37].

The clinician is aware that standards seem to differ and are not identical for competence to refuse treatment and competence to consent to treatment. Should it be said that a person is equally competent to consent to or refuse treatment, and that one should honor (or not honor) the person's decision whether he refuses or consents? Some, like Meisel and Roth, defend the position that there should be symmetry in standards of competency for consent and competency for refusal [29]. This position of symmetry, however, stands in contrast to reports from the President's Commission, which states that whether one is competent depends in part on what one decides ([35], [36].)

Mention has already been made of how non-psychiatric physicians often turn to the psychiatrist for 'permission' to proceed with treatment without involving the courts [5]. The psychiatrist's medical and surgical colleagues call upon him to assess, in medical settings, patients whose status in regard to competency is in doubt. One does wonder if this expectation of 'permission' exerts a special influence on the psychiatrist and colors his decision about the patient's competence.

At some level the psychiatrist is aware that his perceptions of the legal system may be influencing him more than his clinical judgment should permit [16]. When the psychiatrist interacts with the legal system, there is the constant danger that he will abandon the uncertainties of the clinical perspective for the alluring rationality of legal thought. The psychiatrist generally perceives the law as addressing competency as a fixed attribute of a person, a characteristic in itself with a fixed stability. Thus, the psychiatrist may be led incorrectly to assess a patient's functioning at a single time, in a single setting, with an uncertain factual base, and then draw a global conclusion about a patient's functioning [5]. Why is the psychiatrist seduced by that approach when he knows that what the law calls competency is, in fact, a set of deductions from a variety of clinical data that can be as subject to influence and change as the more basic mental attributes on which it is based?

Possibly the psychiatrist should look for positive dimensions in his interaction with the legal system [25]. When psychological incompetence has been established and judicial consent is sought in order to proceed with treatment that is deemed imperative, the court (or law) may serve as a specific directive for conversation. The law gives the patient or his representative and the physician a vocabulary by which to deliberate together and alone, without dictating the script or delivering a monologue of its own devising. This vocabulary proffers the ideas of mental normality, of human capacity for choice, and of individual dignity ([9], pp. 122—123). Thus, the judicial hearing that so often is perceived as a negative, a waste of time and money, or even as a method to rubber-stamp what the doctors are requesting, may turn out to have a salutary and nurturant role in the interaction between doctor and patient.

Louisiana State University School of Medicine
New Orleans, Louisiana

BIBLIOGRAPHY

1. Abernethy, V.: 1984, 'Compassion, Control, and Decisions about Competency', *American Journal of Psychiatry* **141**, 53—58.
2. Ackerman, T. F.: 1982, 'Why Doctors Should Intervene', *The Hastings Center Report* **12**(4), 14—17.
3. Annas, G. J. and Densberger, J. E.: 1984, 'Competence to Refuse Medical Treatment: Autonomy vs. Paternalism', *Toledo Law Review* **15**, 561—596.
4. Appelbaum, P. S. *et al.*: 1981, 'Empirical Assessment of Competency to Consent to Psychiatric Hospitalization', *American Journal of Psychiatry* **138**, 1170—1176.
5. Appelbaum, P. S. and Roth, L. H.: 1981, 'Clinical Issues in the Assessment of Competency', *American Journal of Psychiatry* **138**, 1462—1467.
6. Appelbaum, P. S. and Roth, L. H.: 1982, 'Competency to Consent to Research', *Archives of General Psychiatry* **39**, 951—958.
7. Appelbaum, P. S. and Roth, L. H.: 1983, 'Patients Who Refuse Treatment in Medical Hospitals', *Journal of the American Medical Association* **250**, 1296—1301.
8. Backer, B. *et al.*: 1982, *Death and Dying — Individuals and Institutions*, John Wiley and Sons, New York, pp. 180—182.
9. Burt, R. A.: 1979, *Taking Care of Strangers: The Rule of Law in Doctor-Patient Relations*, Free Press, New York, pp. 122—123.
10. Cahn, C. H.: 1980, 'Consent in Psychiatry', *Canadian Journal of Psychiatry* **25**, 78—85.
11. Cassell, E. J.: 1977, 'The Function of Medicine', *The Hastings Center Report* **7**(6), 16—19.
12. Culver, C. M. *et al.*: 1980, 'ECT and Special Problems of Informed Consent', *American Journal of Psychiatry* **137**, 586—591.
13. Culver, C. M. and Gert, B.: 1982, *Philosophy in Medicine: Conceptual and Ethical Issues in Medicine and Psychiatry*, Oxford University Press, New York, pp. 42—63.
14. Dworkin, G.: 1976, 'Autonomy and Behavior Control', *The Hastings Center Report* **6**(1), 23—28.
15. Dworkin, G.: 1972, 'Paternalism', *Monist* **56**, 64—65.
16. Feldman, W. S.: 1978, 'Is Informed Consent Possible in Psychiatry?' *Legal Aspects of Medical Practice* **6**, 29—31.
17. Friedman, P. R.: 1975, 'Legal Regulation of Applied Behavior Analysis in Mental Institutions and Prisons', *Arizona Law Review* **17**, 39—104.
18. Gert, B.: 1975, *The Moral Rules*, Harper and Row, New York, pp. 27—37.
19. Gutheil, T. G. and Appelbaum, P. S.: 1982, *Clinical Handbook of Psychiatry and the Law*, McGraw-Hill, New York, pp. 210—252.
20. Halleck, S.: 1980, *Law in the Practice of Psychiatry*, Plenum, New York, pp. 268—276.
21. Jackson, D. L. and Youngner, S.: 1979, 'Patient Autonomy and Death with Dignity', *New England Journal of Medicine* **301**, 404—408.
22. Janicak, P. G. and Bonavich, P. R.: 1980, 'The Borderland of Autonomy: Medical-Legal Criteria for Capacity to Consent', *The Journal of Psychiatry and the Law* **8**, 361—387.

23. Kant, I.: 1938, *Fundamental Principles of the Metaphysics of Ethics*, O. Manthey-Zorn (trans.), D. Appleton-Century, New York, p. 65.
24. Katz, J.: 1981, 'Disclosure and Consent in Psychiatric Practice: Mission Impossible?', in C. K. Hofling (ed.), *Law and Ethics and the Practice of Psychiatry*, Brunner-Mazel, New York, pp. 91—117.
25. Kaufman, C. L. *et al.*: 1981, 'Informed Consent and Patient Decisionmaking — The Reasoning of Law and Psychiatry', *International Journal of Law and Psychiatry* **4**, 345—361.
26. Macklin, R.: 1977, 'Consent, Coercion, and Conflicts of Rights', *Perspectives in Biology and Medicine* **20**, 360—371.
27. MacKinnon, R. A.: 1980, 'Psychiatric History and Mental Status Examination', in A. M. Freedman *et al.* (eds.), *Comprehensive Textbook of Psychiatry-III*, 3rd Edition, Williams and Wilkins, Baltimore, pp. 906—920.
28. McCormick, R. A.: 1981, *How Brave a New World? Dilemmas in Bioethics*, Doubleday, Garden City.
29. Meisel, A. and Roth, L. H.: 1981, 'What We Do and Do Not Know about Informed Consent', *Journal of the American Medical Association* **246**, 2473—2477.
30. Meisel, A. *et al.*: 1977, 'Toward a Model of the Legal Doctrine of Informed Consent', *American Journal of Psychiatry* **134**, 285—289.
31. Mill, J. S.: 1962, *Utilitarianism, Liberty and Republican Government*, A. D. Lindsay (ed.), J. M. Dent and Sons, London, p. 73.
32. Miller, B. L.: 1981, 'Autonomy and the Refusal of Lifesaving Treatment', *The Hastings Center Report* **11** (4), 22—28.
33. Morris, G. H.: 1978, 'Conservatorship for the Gravely Disabled: California's Nondeclaration of Nonindependence', *International Journal of Law and Psychiatry* **1**, 395—426.
34. Parsons, T.: 1951, *The Social System*, Free Press, Glencoe, pp. 436—439.
35. President's Commission for the Study of Ethical Problems in Medicine and Biomedical and Behavioral Research: 1983, *Deciding to Forego Life Sustaining Treatment — Ethical, Medical and Legal Issues in Treatment Decisions*, U.S. Government Printing Office, Washington, D.C.
36. President's Commission for the Study of Ethical Problems in Medicine and Biomedical and Behavioral Research: 1982, *Making Health Care Decisions — The Ethical and Legal Implications of Informed Consent in the Patient-Practitioner Relationship*, Volume 1, U.S. Government Printing Office, Washington, D.C.
37. Roth, L. H.: 1981, 'A Commitment Law for Patients, Doctors, and Lawyers', in M. D. Hiller (ed.), *Medical Ethics and the Law — Implications for Public Policy*, Ballinger, Cambridge, pp. 267—281.
38. Roth, L. H. *et al.*: 1982, 'The Dilemma of Denial in the Assessment of Competency to Refuse Treatment', *American Journal of Psychiatry* **139**, 910—913.
39. Roth, L. H. *et al.*: 1977, 'Tests of Competency to Consent to Treatment', *American Journal of Psychiatry* **134**, 279—284.
40. Strub, R. L. and Black, F. W.: 1977, *The Mental Status Examination in Neurology*, F. A. Davis, Philadelphia.
41. Veatch, R. M.: 1977, *Case Studies in Medical Ethics*, Harvard University Press, Cambridge, pp. 42—48.
42. Watson, A. S.: 1984, 'Comment', *American Journal of Psychiatry* **141**, 58—60.

EDMUND D. PELLEGRINO

INFORMAL JUDGMENTS OF COMPETENCE AND INCOMPETENCE

INTRODUCTION

Every clinical decision involves some assessment of the patient's competence — his capacity to make conscious and reasoned choices in matters relating to his own medical and health care. Competence is central to the moral validity of clinical decisions because competence is a necessary prerequisite to informed consent. Incompetence is a moral warrant for others to take actions they perceive to be in the patient's best interests even against the patient's expressed wishes. Precise assessments of competence are essential if patient autonomy is to be fully respected.

Informal judgments of competence are those made by non-psychiatrists as opposed to formal judgments which involve psychiatric or legal procedures. Informal assessments of competence are a daily responsibility of every physician. They are far more numerous than formal assessments made by psychiatrists, psychologists, judges, or courts. Indeed, they are the crucial and the inescapable gateway to formal assessment since only when competence is in doubt are the more formal procedures invoked.

Informal assessments do not differ in substance from formal assessments. They must address the same central question: Does *this* patient possess the capacity to make a conscious and reasoned choice at *this* time, in *this* context, about *this* treatment? Informal assessments are, therefore, subject to the same ambiguity, complexity, possibility of error, and bias, as formal assessments.

Informal assessments, in addition, have several characteristics that give them a special moral urgency. They are usually made privately, by physicians, and are not subjected to the usual scrutiny of formal determinations. Also, they are often made in situations of clinical urgency when neither physicians nor patients enjoy the advantages of deliberate and careful reflection. Finally, they are made by physicians not specifically trained to evaluate disorders of mentation. It is the deliberateness, explicitness, and expertise employed that separate formal from informal assessments.

The importance of competency assessment, its relationships with autonomy and paternalism, and with law and psychiatry have generated a sizeable literature which this essay will not attempt to analyze critically. Rather, my objective is to examine the conditions under which informal assessments are made, the way determinations of competence and incompetence relate to those conditions, and the moral guidelines that the general clinician should follow in making informal assessments. To attain these objectives, we must examine some operating definitions of competence, the impediments to its operation, and the conditions under which informal assessments can be made correctly, objectively, and with as much moral sensitivity as possible.

II. THE COMPETENT PATIENT AND THE COMPETENT DECISION

The most widely accepted definition of competence identifies it as a capacity for performance of a specific task. In medical affairs, the task in question is the limited one of making a reasoned choice among alternative courses of action presumed to benefit a sick person. Clinical competence is a more limited concept than legal competence which invokes the capacity to enter contracts, sign checks, or transfer property.

A distinction can be made between a competent person and a competent decision. (See the more detailed discussion of this distinction by Beauchamp [1] and Morreim [9] in this volume.) However, this essay argues that the exigencies of clinical decision-making are best served by primary emphasis on the characteristics of a competent person, and only secondarily on the 'competence' or 'incompetence' of the decision itself.

A competent person possesses the capacity to make an explicit, reasoned, and intentional choice among alternative courses of action. At a minimum, this capacity includes some substantial measure of the following capabilities: (1) The capacity to receive, comprehend, retain, and recall information provided by others or gathered by oneself. Information may be by word of mouth, reading, or some set of mutually understood set of signals. (2) The capacity to perceive the relationship of the information received to one's current clinical predicament as a sick person for whom something might or might not be done to cure, contain, or prevent illness or its symptoms. (3) The capacity to integrate, order, and relate the information received to a realistic perception of the need for making a choice in such a way that the patient can weigh benefits and risks against some set of personal values. (4) The capacity to select an option, to give cogent reasons for the choice and to persevere in that choice, at least until the decision is acted upon. (5) The capacity to communicate one's choice to others in an unequivocal manner. Among these 'capacities', I would include Gert and Culver's requirement that the patient understands the activity in question and knows when he is doing it [5]. This is, I believe, implicit in the use of the terms 'conscious and reasoned choice' which I employ in this essay.

A person possessing these capabilities should, other things being equal, be competent. A person lacking one or another of them is not necessarily incompetent. Whether competence should be a yes-no decision or whether competence exists on continuum, is an arguable question. Wikler suggests two options: one is that competence is a matter of degree, that people possess more or less competence, and that the dividing line is a thin and arbitrary one. The other option is that there is a threshold for competence, and, once over that threshold, people will be equally competent to perform a given task, though their additional capacities beyond the threshold may differ substantially. ([18], pp. 377–392).

Beauchamp and McCullough suggest that Wikler's dichotomy can be eliminated if "equally competent" is replaced by "sufficiently competent" ([2], pp. 118–128). This seems a reasonable compromise since the empirical question is whether, in this instance, the patient is capable of a reasoned and conscious choice, not whether he can make the decision superlatively well.

One caveat, to which I shall refer later, is to avoid tying competence so closely to the nature of the decision that anyone not making the 'right' decision in a given situation is considered insufficiently competent. Beauchamp and McCullough tend in this direction in their discussion of the concept of competence and the criteria for intervention [2]. This distinction is important because, in its absence, there is the danger of conflating the competence of the patient and the competence of his decision.

Whether one accepts the threshold or the relativist conceptions of competence, the empirical assessment of competence depends upon criteria or tests which provide no absolute dividing line between competence and incompetence. The criteria used to measure competence and incompetence are external evidences of an interior capacity for conscious and reasoned decision and choice. The crucial empirical question is where to draw the threshold line or point on a continuous spectrum from unquestioned competence to unquestioned incompetence.

The criteria themselves, their validity, and their relationships to the criteria for autonomy are treated in more detail in the other papers in this volume (Morreim [9], Beauchamp [1], Pincoffs [12], Knight [8]) and need not be repeated here. What is important here is that judgments of incompetence or competence are, in fact, made on the basis of possession of some threshold combination of capacities which are ascertainable with varying degrees of certitude.

Possession of the threshold level of capabilities for competence does not mean they will, or necessarily can, be exercised. Various forms of coercion or duress, overt or covert, as well as erroneous information, fear, anxiety, the pathophysiological accompaniments of illness — all can impede the operation of the capacities for competent judgments. But automatic declarations of incompetence based on the mere existence of one or more of these factors is unwarranted.

Possession of the capacities for competent judgment does not by itself assure informed consent. Informed consent implies competence, but also autonomy, and adequate information on which to make a choice. On the other hand, adequate information and autonomy do not assure competence. There must also be the capacity to process the information and to use freedom in a reasoned way.

Competence and autonomy are closely allied and the criteria for the determination of each are closely inter-related. Suffice it to say for the purposes of this essay that they are not identical concepts. Incompetent patients may act autonomously — without external coercion, manipulation or deception. Some competent patients, even under severe external duress, may retain their capacities for competent judgment and decision; others may not. Autonomy focuses on self-governance, the freedom to make one's own decisions; competence focuses on making one's own decision in a reasoned and conscious way. Internal factors like great fear, neurotic obsessions or compulsions, or drug addiction, can impair both autonomy and competence.

Even in ordinary decisions, autonomy and competence are relative, and their operation difficult to evaluate. Most significant human decisions — to buy a house, to marry or divorce, or to take a new job — involve some degree of anxiety, fear, group or familial pressures, or threats to property, status, or finances. One sign of the capacity for competence is the ability to appreciate how these factors operate, and how they should be weighed in making a decision.

Even in the absence of illness, we recognize that competence is not a global or a permanent capacity. We recognize that we are competent for some tasks and not for others — and that at different stages of our lives, we possess and lose our capacity for competence for some specific task. In the presence of serious illness, this propensity of selective intermittency of competence is exaggerated. The patient may be confused, disoriented, and senile, yet be able to grasp the meanings and necessity of a limited immediate medical decision while lacking competence for larger decisions like management of property or investments. Even the psychotic patient may be competent to make decisions about treatment, provided they do not involve his delusional system.

Competence is a capacity; the decision is the product of that capacity. One need not necessarily follow the other. Ordinarily, a patient capable of making conscious and reasoned choices can be expected to make a competent decision. But the 'competence' of the decision is not synonymous with the 'competence' of the patient. The decision may be considered erroneous or harmful when it is not in conformity with the physician's recommendations, with some external calculus of risks and benefits, or with the decision a 'reasonable' person would make. But that does not, *ipso facto*, make the patient incompetent. On the other hand, a totally incompetent patient could make the 'right' decision, i.e., by agreeing with the medical recommendation, without the capacity for consciously reasoned choices.

When a competent patient makes a decision that may bring harm or death, that decision is usually the result of differences in the values or factual premises upon which it is based, or an error in logic. For example, the patient may be given insufficient or erroneous information about a procedure or its risks and benefits. More usually, the judgment that an otherwise competent patient has made an 'incompetent' decision arises from differences in presuppositions about fundamental values of a religious or philosophical nature that, in the patient's life, take precedence over all other considerations. We might argue that the patient's values are distorted, that we cannot in good conscience carry out his wishes, or that we ought, on the principle of beneficence, to intervene against the patient's choice. But disagreements at this level cannot justify labeling the patient as incompetent, unless there is evidence of a defect in the capacity to make a reasoned decision.

The same applies to self-imposed constraints on freedom of choice. Roman Catholics are not free to participate in abortion or direct euthanasia; Orthodox Jews are not free directly to shorten a dying patient's life; a Jehovah's Witness is not free to accept a transfusion. Most humans have some set of self-imposed constraints which cannot legitimately be counted as evidence against either competence or autonomy. When the premises are empirically falsifiable, as with some psychotic patients, then they do constitute impediments to competent decision-making.

Competence in the medical setting is the capacity for conscious and reasoned choices about one's own health and medical care. As a concept, it is limited to the specific clinical instance. It is not global; it need not be permanent; it is judged by the capacity to make a consciously reasoned decision and not necessarily by the content of that decision. It is empirically tested by criteria based on an assessment of the external signs of the operation of the internal capacity to make reasoned choices. These criteria measure the possession of some threshold level of capability, beyond which a person is presumed to be sufficiently able to make his own decision, and to be permitted to do so. Thus, even if we prefer the concept of competence as a continuum, we end in a yes/no decision. The clinician proceeds with a procedure or does not.

III. COMPETENCY ASSESSMENTS IN ACUTELY ILL PATIENTS

All the uncertainties in the concept of competence and its empirical evaluation are compounded in acute illnesses and emergencies. In these circumstances, physicians are confronted with urgent situations, in which decisions must often be made quickly, often without adequate information or preparation of the patient or the family. In addition, the illness or injury may be attended by pathophysiological disorders of mentation. Such disorders may result from pathological lesions in the brain itself — such as Alzheimer's disease, HIV infection, tumors, encephalitis, stroke, meningitis, brain contusion, or laceration. Or, the primary source of disordered mentation may arise outside the brain, affecting its function temporarily — e.g., anxiety, pain, shock, medications of all sorts, dehydration, azotemia, acidemia, electrolyte disturbances, hypoglycemia, anoxemia, or hypercapnia. These latter are often accompaniments of treatable disorders. They may impair brain function temporarily, but can often be reversed by appropriate therapeutic maneuvers.

Under these circumstances, it is hardly imaginable that all of the detailed criteria for assuring competent decisions can be fulfilled. The physician is caught between his duty of beneficence, on the one hand, and respect for the patient's choice, on the other. How competent is a severely burned patient who asks not to be treated? [17], pp. 9—10). How competent is a patient with airway obstruction which causes severe anoxemia, hypercapnia, and acidemia? Or the patient with increased intracranial pressure who refuses or, for that matter, accepts craniotomy? Or the patient with the fluctuating mention of Alzheimer's disease?

Matters are further complicated when the acute episode is part of a chronic, recurrent, ultimately fatal disorder. One might presume in favor of overriding the patient's choice not to be treated in a first acute episode when the doctor knows little about the patient's values. The patient may recover from the first, or subsequent episodes, and then make a competent decision not to be treated. He may change his mind again when a new acute episode occurs. Patients, for example, may ask, when not in acute distress, that vigorous resuscitative measures not be used in their next episode. The decision may be made with doctor and family assenting and incorporated into a living will. The decision might fit the criteria for competency. Yet, when the actual moment of decision arrives, and the patient is bleeding acutely, or unable to breathe, he may well ask that his decision be reversed.

There is usually little time in emergencies or acute illness for formal determinations of competence. Often the physician is forced to make the roughest kind of assessment. In addition to the inherent uncertainties, the decision may be influenced by the duress experienced by the physician. Decisions in acute emergencies are attended by urgency, confusion, and the distress of families and friends whose advice and information about the patient's wishes may be contradictory or inadequate. The physician may fear a malpractice suit; he may be an inexperienced resident.

Not all physicians are endowed with equanimity or objectivity that assures them of a capacity for conscious and reasoned judgments. The physician's duress may undermine his own capacity for conscious reasoned judgment in two ways: assessing the patient's competency or making the technically correct decision. This is a factor that deserves some attention since informal assessments of competence must sooner or later take into account the physician's psychological competence as well as the patient's.

Several clinicians have examined the special difficulties of assessments of both autonomy and competence in the actual clinical situation. [7], p. 301; [15], pp. 12–15). They underscore the factors that make these judgments difficult and uncertain, that influence the physician's decisions, and that diminish the patient's capacities for reasoned choice. Their central concern is for the conflict between the physician's traditional commitment to beneficence and non-maleficence and the requirement to respect patient autonomy. Jackson and Youngner, [7] and

Siegler [10], like Childress, tie assessments of competence closely to what should be done in treating the incompetent patient ([3], pp. 63–65, 103–107). While this is an important relationship, tying incompetence and intervention too closely slants the assessment of competence rather strongly in the direction of intervention.

Siegler [15], for example, outlines six factors that in his judgment influence the physician's decision in critically ill patients about respecting the patient's wishes: the patient's capacity to make a choice, the consistency of the choice with the patient's prior values, the patient's age, the treatability of the disease, the physician's value system, and whether the decision is being made in the hospital, the doctor's office, or patient's home. Of these six criteria, only the first deals specifically with the patient's competence. The others concentrate on whether the decision the patient has made should be respected.

Age is significant only if the patient is too young to be competent. But age is a relative matter since children mature at different rates, physically, intellectually, and psychologically. Many children under the legal age for independence can make competent decisions while not a few over cannot do so.

Consistency of the choice with the patient's prior or 'true' value system is not always relevant to competence assessment in the acute situation. Under such conditions, the patient's values expressed in prior assertions about what he would want done could change drastically. In the face of the immediate consequences of an earlier decision, the change may be in either direction, for or against treatment. Indeed, the patient may change his values several times so that instability of values is not *prima facie* evidence of incompetence. The psychological stress and uncertainties of acute illnesses make vacillation and velleity the normal accompaniments of patient choices.

For similar reasons, Jackson and Youngner's identification of 'true' autonomy with the choice the patient would have made were he in a calm and deliberate state is dubious. ([5], [15], pp. 21–22). The patient is *de facto* not in a calm or deliberate state when he is making his choice. Few of us can predict what choices we, or others, might make when we actually experience the pain, discomfort, anxiety, and prospects of death an acute illness may present. These factors are part of the decision in acute illness. To take 'true' autonomy out of the actual context of the decision is to abstract it from reality. Indeed, to change one's mind in the face of the actual decision may be a truer sign of true autonomy or competence than consistency with a theoretical choice made in the past.

Some physicians take the state of illness itself to be an automatic impediment to competent decisions. They argue that the dependence of the sick person, the difficulties of understanding technical terms, and the need to trust the physician's judgment make competence and informed consent moot points. Some even assert, without compunction, that they can get any decision they want by the way they present the 'facts'. They assume that all patients are, in part, incompetent and that the physician therefore has a duty to manipulate the decision in the 'right' direction.

The phenomenology of illness does, indeed, impose limitations on the patient's freedom in several domains. The sick person is forced to accommodate to the demands of pain, disability, and anxiety. His body becomes a central focus for conscious concern and a preoccupation. The ill person is forced to depend on the knowledge, skill, and integrity of the physician. The physician can exert considerable power for good and evil. These facts make the relationship one of inequality, and the sick person is often forced by illness into a vulnerable and exploitable state which is highly variable from person to person. ([10], pp. 70—79)

None of this justifies an automatic judgment of incompetence to make decisions. All decisions, clinical and otherwise, occur at bifurcation points in life and imply some degree of stress. The presence of obstacles to competence imposes a special obligation to remove them in order to enhance competence to the extent possible, and, thus, to restore the locus of decision-making to the patient. To assume incompetence, or to place obstacles in the way of the patient's exercise of his capabilities for competent judgment (e.g., by deception or incomplete information) is a serious misconstrual of beneficence.

The treatability of illness, even its gravity, should not predetermine the assessment of competence since the question is whether the patient wants treatment, not whether he needs it. Similarly, the place in which the decision is made — home, office, or hospital — may condition how the decision about competence is made, but does not affect the patient's competence. Some physicians say that competence is an illusion since they are capable of getting any decision they want by deception or subtle coercion. Many will justify doing so for the "good" of the patient. They interpret fidelity to trust as assuring the medically good decision, not protecting autonomy. This is morally indefensible. The patient's vulnerability and dependence only magnify the physician's obligation to obtain a morally valid decision, which *ipso facto* means a decision which is autonomously and competently made.

The physician's own values are relevant only if the decision of a competent patient conflicts with the physician's conscience, and he feels he cannot cooperate with it. As already pointed out, disagreement with the physician's recommendations is not a final criterion of competence.

Siegler's criteria are an accurate reflection of the criteria physicians actually use in decisions to intervene more than in decisions about competence. There is some justification for their use on the grounds of clinical discretion ([15], pp. 12, 28–29). But clinical discretion does not vitiate the importance of keeping decisions about the patient's competence separate from decisions about interventions against his will. The patient's best interests, as a rational and sentient being, are better served by making these judgments in separate steps, related to, but not telescoped into, one another.

IV. AUTONOMY AND BENEFICENCE IN COMPETENCE ASSESSMENT

Despite the physician's best efforts to make competency assessment as objective as possible, value desiderata will intrude themselves. They cannot be eliminated as long as medicine remains an interaction of persons. Beauchamp and McCullough underscore how strongly value presuppositions may condition the stringency with which physicians apply the criteria of competence. [2] They call for clearer reasoning to deal with the potential conflicts that may arise between the principles of beneficence and autonomy at the moment of clinical decision.

This is certainly a necessary step. But reasoning alone will not resolve deeper conflicts of values, since they arise in a conflict of axioms and not logic. The key question is: which should be the dominant principle — the patient's freedom to choose what he thinks is good for him, or the doctor's freedom to intervene when, in his opinion, the patient has made a harmful or dangerous choice? Roth and his co-workers try to resolve this matter by relating their criteria of competence to the gravity and consequences of the decision to be made. When the decision involves a treatment that could save the patient's life — for example, transfusion in severe blood loss — they would recommend that the more stringent criteria for competence be applied. When the decision is of a less definite nature, for example, treating arthritis, they allow for a much looser interpretation. [13].

This solution has several shortcomings. First of all, it reinforces the relativistic character of competence assessment. It increases the likelihood of fitting competency assessment to the physician's values and puts the patient's autonomy at considerable risk. It gives too easy an automatic warrant for medical intervention over the patient's wishes.

Another shortcoming is the presumption that beneficence is synonymous with strong or weak paternalism. But beneficence can also be interpreted as acting in such a way as to enhance the patient's autonomy since autonomy is, itself, one level of good [11]. On this view, we do harm to the patient if we override patient autonomy by declaring him incompetent in the really important decisions.

These considerations take us back inevitably to the need for clarification of our conceptions of beneficence, autonomy, and the patient's good. These crucial questions are obfuscated by assessments of competence that are situation-dependent. Such assessments beg the question by presuming that when the patient disagrees with the doctor, beneficence requires that autonomy be violated.

It seems preferable to consider competence as a capacity of the patient to be judged independently of the gravity of the decision, of the doctor's recommendations, or what a reasonable person would choose. The first step is to determine, as objectively as possible, whether the patient is competent. If the patient is judged incompetent, then we must face the next question: What moral warrant do we have for intervention against his choice?

This approach forces us to think more rigorously about imposing a choice on a competent patient. We cannot hide the decision behind a situation-dependent scale of competence. The light of argument must be, therefore, focused where it belongs, — on the conflict between fundamental values of beneficence and autonomy, and on the need for a more holistic view of the good of the patient [11].

There is one aspect of intervention that is so closely tied to competence determination that it cannot be neatly separated from it. I refer to interventions designed to remove remediable causes of incompetence like ventilatory therapy to treat anoxia or hypercapnia, restoration of fluid and electrolyte balance, dialysis of azotemic patients, discontinuance of toxic drugs, or treatment of infection. Under these circumstances, the physician is acting to restore competence so that the patient may participate in decisions that affect his well-being. In so doing, he is acting paternalistically, but in a morally defensible way.

Beauchamp and McCullough propose a graded system of justifications for intervention based upon the degree of reversibility of competence (and autonomy) [2]. They give autonomy precedence when competence is restorable, and when a 'value history' is available. They give precedence to beneficence when autonomy is irreversibly lost. Their system would accord roughly with the hierarchical ordering of patient good which I have proposed elsewhere.

I disagree, somewhat, with Beauchamp and McCullough's contention that treatments aimed at restoring autonomy or competence should not be undertaken in the incompetent patient if he gave prior instructions against such treatments [2]. My disagreement is limited to the severely and acutely ill patient. Empirically, we find all too often with such patients that a prior request against specific treatments is often reversed when the full exigencies of a rejection are fully comprehended. The patient is owed restoration of autonomy. That means that only those measures aimed at relief of pathophysiological causes of incompetence should be used, not treatments of the underlying illness if the patient has opted against them.

Beauchamp and McCullough, themselves, seem to allow for some relaxation of their rule of freedom over beneficence in acutely ill patients. These are just the situations we have concentrated on in this paper — those in which the opportunity for a "relaxed philosophical inquiry", as they put it, is impossible ([2], p. 132).

V. THE COMPETENCE OF PROXIES, SURROGATES, AND GUARDIANS

When a patient is judged incompetent, the physician usually turns to some surrogate or proxy, like a court-appointed guardian, the family or friends, to act in the patient's behalf. This is to assure that the patient's values and wishes are taken properly into consideration, and that as much information as possible about these values will be factored into the decision. When dealing with proxy consent, the physician retains his prime responsibility to act in the patient's behalf. He has an obligation to assess the competency of the proxy since he depends on him for several important functions: (1) providing information not otherwise accessible about the patient's wishes and values, (2) making a reasoned choice in the patient's stead, and (3) evaluating the patient's past behavior which enters into the assessment of the patient's present competence.

These three functions require a proxy who exhibits the same criteria of competence required of the patient, were the patient making the decision. The physician's prime relationship is with the patient even if he never knew the patient in a competent state. He, as well as the proxy, must act in the patient's behalf and must, in a sense, judge each other's capacity to do so since each has an obligation to do the patient's good.

We cannot assume that a member of the family, a friend, or a court-appointed guardian will necessarily be able to speak more authoritatively in the patient's behalf than the physician. Family members and friends will ordinarily know more than the physician about what the patient himself might have wanted or wished, but they may not be competent to relate this to the clinical situation. Moreover, they may be under such severe emotional stress, or so impaired in their own reasoning capabilities, that they cannot make a competent decision for someone else.

It often happens, too, that parents and friends appear during an acute illness after not having seen the patient for many years. They may, nonetheless, presume to speak for the patient or to offer observations on whether he is in his 'right mind'. The justification for proxy decisions rests on the high probability that family members or friends will more carefully protect the patient's interest than non-family members. But this probability does not confer absolute moral authority, and it may not turn out to be a reality at all.

Even when the probability is high, it does not follow that the proxy will act in the patient's best interests. There may be some conflict of material interest — a will, property, long-standing family feud, or animosity — that might compromise the proxy's capacity to receive, process, reason about, or transmit information helpful to the patient or to a decision in his best interests.

Some judgment about the competence of proxies of all types seems essential if the patient is to be protected against unjustifiable encroachments on his autonomy. The physician cannot abandon his patient automatically to the will of the proxy without failing in his duty of fidelity.

VI. SOME MORAL GUIDELINES FOR INFORMAL ASSESSMENTS OF COMPETENCE

Granting the difficulties of informal assessments of competence, there is need for some guidelines for making morally defensible competence assessments. To propose a set of guidelines requires some statement of the antecedent ordering principles on which they are based. Space will not permit a defense of these principles and they must be set forth here without necessary argumentation.

I would propose that the ordering principle be the principle of beneficence, so interpreted that preservation of autonomy becomes a component of the patient's good and therefore consistent with, and not in opposition to, benevolence and beneficence [11].

Even in complex and urgent clinical situations, therefore, the clinician should adhere to some explicit set of criteria of competence to the degree the clinical exigencies permit. He has a duty, therefore, to become particularly skillful in informal assessments of the disorders of mentation that may signal incompetence.

This is, incidentally, a neglected aspect of physicians' training in which the usual assumption is that medical good overrides all other good in the care of seriously ill patients. As guardians of the gateway to more formal assessment, general clinicians must be able to evaluate the possibility that the patient may not be competent to make reasoned choices. The physician should then seek formal assessment through consultations. If there is not time for a more formal assessment of psychologic competence, the clinician must act so that whatever competence the patient possesses is utilized or enhanced in making decisions.

A particular responsibility of clinician is to evaluate as objectively as possible all those pathophysiologic or pathological conditions that can affect brain function and mentation, and therefore the capabilities needed for competent decisions. He and his colleagues need this information when making informal assessments, and so does the psychiatric consultant when making a formal assessment. The clinician must also ascertain which of the impediments to competent decision-making are reversible and to what degree, and which are not.

The clinician has an obligation to enhance whatever degree of competence is present by treating reversible disorders of cerebral function [2]. When this is accomplished, competence must be reassessed and the patient's choices restored to preeminence in decision-making. In treating reversible pathophysiological disturbances, the physician may override the objections of the patient unless the patient before becoming incompetent clearly, explicitly and specifically asked that such measures not be used. But even in these instances, if there is any doubt at all, competence should be restored to allow for as autonomous, and competent, a set of subsequent decisions as possible.

When patients wax and wane in their competence, and vacillate in their decisions for or against treatment, the last substantially competent decision should prevail. This will often be the case in chronic illness marked by acute exacerbations. Velleity and vacillation are not warrants for an automatic diagnosis of incompetence.

The physician must provide the patient or his proxy with the information necessary to make an informed and competent choice. Suffice it to say that the information should be provided without deception, coercion, or manipulation. The information also must be presented sensitively, in a way geared to the patient's capacity for understanding, to his education, and to his linguistic, cultural, and social milieu. This is an element in competent decision-making very much under physician control. For that reason, the physician has a moral obligation to take special pains in informing both the patient and the family. He must avoid the all-too-frequent judgment that patients cannot comprehend the choices because they must be couched in technical language.

When dealing with the proxy consent of family, friend, or legal guardian, the physician has an obligation to satisfy himself that the proxy is competent to make a choice for this patient, and that no conflict of interest with the good of the patient exists. If the proxy choice conflicts with the good of the patient, the physician's first obligation is to his patient. He is not bound automatically to accept a proxy decision.

When the physician finds the choice of a competent patient, or of the proxy, offensive to his own moral principles, he should offer to withdraw from the case without endangering the patient's clinical outcome. He need not seek out another physician who will agree with the patient, if he believes that the patient's choices involve a serious breach of morality (for some physicians suicide, or abortion). It is the responsibility of the competent patient, or his proxy, to inform the next physician of the patient's values and preferences before entering the new therapeutic relationship. During these discussions, the original physicians must continue to care for the patient to avoid the change of abandonment.

In all of this, the assessment of competence should be made as objectively as possible, and distinct from the gravity of the decisions to be made, and how to deal with the patient judged to be incompetent. These domains obviously must be related to each other, but in stepwise and modular fashion, not by blurring the distinctions among them.

Some of the difficulties of a sliding scale of criteria for competence are evident in Drane's proposal of three standards of stringency fitted to the gravity of the choice. Drane proposes that the least stringent criteria — awareness and assent — should be used with non-dangerous treatments that are in the patient's best interests. A more stringent criterion — understanding and ability to choose — are required if one's illness is chronic, treatment more dangerous and of less certain benefit. The most stringent criterion — appreciation of consequences — is to be used if refusal of treatment would mean death.

Drane's proposal has a certain pragmatic attractiveness but it suffers from several difficulties. Its author justifies his procedure by saying that it balances values. But this is perhaps its deficiency, too, since it is the physician who decides what degree of stringency should be used. Thus competency becomes a manipulable concept, consciously or unconsciously. Moreover, this approach measures the rationality of the decision by conformity with some external standard of reasonableness, confusing the competency of the patient with the competency of his decision.

In effect, the patient is allowed to make the non-significant decision but is considered progressively incompetent as the significance of the decision increases. Instead of starting with a presumption of competence, Drane's proposal 'ratchets up' the criteria so that the presumption is that the patient is incompetent [3].

If the patient's capacity for conscious reasoned choice is truly impaired, it should be so judged, and appropriate steps taken to protect his interests — whether the choice is serious or not. All the difficulties of making competency assessments stressed in this essay are compounded by a shifting definition.

Informal assessments of competence are the everyday business of every physician. They require more explicit formulation than is now customary because competence is so central to informed consent, decisions to intervene in the patient's behalf, and protecting the patient's best interests. Precise, sensitive, and continual informal judgments of competence remain the pivotal points in protection of patient autonomy.

Georgetown University
Washington, DC.

NOTES

[1] Stanley *et al.* [16] provide data to show that elderly patients' choices are as 'reasonable' as those of younger patients, even though their comprehension of consent information is significantly less. Here is an example where the physician's obligation to enhance competence must focus on enhancing comprehension.

[2] Drane [4] calls for a sliding scale of criteria for competency.

BIBLIOGRAPHY

1. Beauchamp, T. L.: 1991, 'Competence', in this volume, pp. 49—77.
2. Beauchamp, T. L. and McCullough, L. B.: 1984, *Medical Ethics: The Moral Responsibilities of Physicians*, Prentice Hall, Englewood Cliffs, New Jersey.
3. Childress, J. F.: 1982, *Who Should Decide Paternalism in Health Care?*, Oxford University Press, New York.
4. Drane, J. F.: 1984, 'Competency to Give Informed Consent', *Journal of the American Medical Association* **252**(7), 925—927.
5. Gert, B. and Culver, C.: 1982, 'Philosophy in Medicine, Conceptual and Ethical Issues', in *Medicine and Psychiatry*, Oxford University Press, New York, Chapter 3.
6. Jackson, D. L. and Youngner, S.: 1980, 'Comment on Family Wishes and Patient Autonomy', *Hastings Center Report* **10**, 44.
7. Jackson, D. L. and Youngner, S.: 1979, 'Patient Autonomy and Death With Dignity', *New England Journal of Medicine* **57**(1), 404—408.
8. Knight, J.: 1991, 'Judging Competence: When the Psychiatrist Need, or Need Not, Be Involved', in this volume, pp. 3—28.
9. Morreim, E. H.: 1991, 'Competence: The Interaction of Law, Medicine, and Philosophy', in this volume, pp. 93—125.
10. Pellegrino, E. D.: 1981, 'Being Ill and Being Healthy: Some Reflections on the Grounding of Medical Morality', *Bulletin of the New York Academy of Medicine* **57**(1).
11. Pellegrino, E. D.: 1983, 'Moral Choice, The Good of the Patient and the Patient's Good', in L. Kopelman and J. Moskop (eds.), *Ethics and Critical Care Medicine*, D. Reidel Publishing Company, Dordrecht, Holland, pp. 117—138.
12. Pincoffs, E.: 1991, 'Judgments of Incompetence and Their Moral Presuppositions', in this volume, pp. 79—89.
13. Roth, L. H. *et al.*: 1977, 'Tests of Competency to Consent to Treatment', *American Journal of Psychiatry* **134**, 279—284.
14. Siegler, M.: 1982, 'Commentary: Does Doing Everything Require CPR?', *Hastings Center Report* **12**.
15. Siegler, M.: 1977, 'Critical Illness: The Limits of Autonomy', *Hastings Center Report* **7**.
16. Stanley, B.: 1984, 'The Elderly Patient and Informed Consent', *Journal of the American Medical Association* **252**(10), 1302—1306.
17. White, R. B. and Engelhardt, H. T.: 1975, 'A Demand to Die', *Hastings Center Report* **5**, 9—10.
18. Wikler, D.: 1979, 'Paternalism and the Mildly Retarded', *Philosophy and Public Affairs* **8**, 377—421.

PART II

REEXAMINING CONCEPTS OF COMPETENCE

TOM L. BEAUCHAMP

COMPETENCE

A number of conceptual and normative assumptions that underlie judgments of competence and incompetence have practical implications for medicine, clinical psychology, law, and public policy. These judgments can trigger paternalistic, coercive, and authorizing events of the most consequential sort for patients. For example, determinations of competence and incompetence can function to distinguish patients whose consents and refusals will be accepted — and treated as binding — from those whose decisions will not be so treated. These determinations thus can serve to open the door to the dismissal of requirements of informed consent, as well as to coercive treatment or confinement.

In this paper, I deal generally with the nature of competence, but with special attention to the following: competence to consent (part I), the relationship between the concepts of competence and autonomy (part II), and a range of normative issues surrounding the selection and use of criteria and tests of incompetence (part III). Throughout, my argument is directed at moral problems of incompetence to consent and refuse and at *conceptual* problems in distinguishing competence and autonomy. While mine are philosophical rather than legal or psychiatric remarks, I make frequent reference to the legal and medical literatures. A major and basic question addressed in these literatures is, "Under what conditions is a person able to make an informed choice, in particular to make an informed consent (or refusal)?" I argue that any adequate answer to this question must supply answers to prior questions about the nature of specific *tasks*, the *abilities* required to perform the designated tasks, and the *functions* of judgments of competence. My first objective is to show why this is so.

I. THE NATURE OF COMPETENCE

A. *The Meaning of 'Competence' and the Criteria of Competence*

Medicine, law, psychiatry, philosophy, and other professions have presented competing theories of the abilities that persons must have to be competent. The word 'competence' has a multi-disciplinary history that has spawned layers of criteria of competence that are connected in diverse ways, often at cross purposes. This situation has led some to observe that there is no standard *definition* of competence ([15], [23]), consequently no accepted *test* of competence ([1], [2], [10], [24]), and consequently no nonarbitrary *line* that can be drawn between those who are competent and those who are incompetent ([17]). But let us keep definitions, tests, and boundary lines as distinct as possible, confining attention for the moment to the problem of definition.

I hold, as have others [6], that 'competence' has a single, basic, skeletal meaning that underlies various criteria of competence: 'X is competent to do Y' always means 'X has the ability to perform task Y'. 'Competence' thus *means* 'the ability to perform a task'. This is the term's simple definition (and its logically necessary and sufficient condition). In stark contrast to this *meaning* of 'competence' are the *criteria* of competence — i.e., the criteria for the correct application of the concept of competence. These criteria are not uniform across contexts, because the criteria vary in accordance with the specified task(s). The criteria for someone's being a competent anesthesiologist are different from the criteria for someone's being a competent banker or a competent mule driver, because the particular tasks involved in these distinct activities are different tasks. Learning the criteria of competence applicable to a new class of items is always a new lesson, but we are able to use the word 'competence' for entirely new classes of tasks that we have never before confronted, because a new lesson in the logic or meaning of the concept is never involved once we understand the concept's general character.

It follows that judgments of competence are impossible unless both a task or set of tasks and criteria for the evaluation of a person's performance are presumed or specified. If someone says "Peter is incompetent", a necessary response is "incompetent to do *what*?" — to control his gambling?, to recognize a dog?, to remember persons and facts?, to decide whether to see a doctor? If the *what* is 'speak French', then speaking French is the task, and established criteria of performance will necessarily govern what counts as a competent linguistic performance. If the *what* is 'give directions', then giving directions is the context, and there are criteria operative in the context governing what counts as giving directions. Thus, depending on the task specified or assumed, one could be competent or incompetent to stand trial, to raise a child, to manage financial affairs, to consent, or to lecture about informed consent. The *meaning* of 'competent', however, is not modified even when categorically different tasks are specified. This task-specific character of competence is the bombshell in the basement of the idea that there could be a single set of criteria of competence, such as decision-making ability, rationality, understanding, or some outcome of a decision process. None of these criteria could be solely sufficient for the range of possible tasks.

The tasks cited thus far are specific tasks, but it is also meaningful to describe persons as *generally* competent, a judgment that assumes as its context the ordinary affairs of life. A distinction should therefore be drawn between *general* competence and *specific* competence. General competence refers to a general ability to perform tasks, whereas specific competence requires a specific context with a set of defined tasks. The criteria of general competence, I shall argue, roughly correspond to criteria of being an autonomous person.

A confusing feature of informed consent settings is that judgments often have to be made either about a broad range of abilities defining competence to consent or about a very specific competence to consent to a particular intervention or medication. Whether a patient or a subject can make an informed choice in a specific instance may require no investigation whatever as to whether he or she is able to make informed choices generally. But sometimes we want to know whether a person has broader competence to consent, because we want to know whether the person qualifies as the *kind* of person from whom it is appropriate to solicit consent at all. For example, this problem often arises in the instance of minors near the age of majority or when the possibility of a court-appointed guardian is under consideration.

We shall return to this distinction between what I call a 'broad competence to consent' and a 'narrow competence to consent' momentarily. But in order to appreciate its importance, we must first discuss the function of the concept of competence.

B. *The Function of the Concept of Competence*

The *function* a concept serves often yields more insight into the nature of the concept than a study of its *meaning* and *criteria*. Competence is an example, because it is used to grade, screen, and otherwise evaluate persons (sometimes even animals and machines). What I shall call the *gatekeeping function* of the concept is absolutely central to its invocation in the contexts of interest to us in this paper. This is not, however, the only evaluative function of the concept. We shall later see that judgments of incompetence are also used in the law to change the status of person's rights and to justify interventions.

Consider the statement "He is a competent surgeon". This assertion is often not reducible in meaning to "He is a surgeon, and he can cut, use sutures, stitch, give orders, and the like." These are abilities that the surgeon as a matter of fact either possesses or does not possess, and they are the criteria on the basis of which we say that a surgeon is competent. Nonetheless, terms describing such factual properties do not exhaust the logic of 'competence' in any given setting, because part of what we do with 'competence' is evaluate persons' skills, rightful authority, and the like.

'Competence' can be applied to patients and to health professionals, and indeed in any context in which tasks are performed and evaluations of persons' abilities are desired. When patients use the word 'competence', they may be asking about, assessing, or grading the professional skills of those who treat them. When health professionals use the word to refer to patients, they generally use it both to assess the patient's ability to make decisions and to certify that the person is entitled to a significant measure of control over decisions about his or her treatment (or not entitled, if a judgment of incompetence is reached). These judgments are intended to ensure that patients possess the decision-making abilities needed for mature, prudent, and responsible decision-making.

Judgments of competence and incompetence serve a gatekeeping function in a consent setting by telling us whether X is the *kind of person* from whom it is appropriate to solicit a consent or whether a guardian should be appointed to look after X's interests. That is, such judgments determine whether the application of informed consent requirements should be triggered or some alternatives invoked, whether a guardian should be appointed to look after X's interests, whether involuntary institutionalization is appropriate, and the like. Competence has a distinctive normative role of qualification or disqualification (one is 'graded' on a pass/fail basis), and the concept itself cannot be understood apart from this role. In informed consent settings, the gatekeeping function is to trigger — or if incompetence is found, to preclude — the application of informed consent requirements. A requirement that a competent patient should be *approached* to give an informed consent thus should be sharply distinguished from informed consent requirements of disclosure, comprehension, voluntariness, and the like. They are entirely different in nature and function.

If a person is judged a competent person, he or she has to be dealt with as his or her *own* person. His or her decisions, and not those of another, should be overriding in such matters as choice and consent. But *incompetent persons* — i.e., persons of general incompetence — do not have the ability to choose on the basis of information, or do not have the ability to choose voluntarily, or lack *some* crucial ability. The will of *another* — a guardian, e.g. — may therefore authoritatively prevail. The incompetent person systematically fails to pass through the gate.

The word 'competence' does not always serve such a gatekeeping function, of course. Judgments that X is competent to roller skate usually neither inform us whether to approach X for something, nor trigger any moral obligations we have in regard to X. The gatekeeper function is present only if the range of abilities constituting a competence is fixed in an institutional, social, or legal context in which passage through a gate is controlled by official gatekeepers. Competence serves a gatekeeping function in a wide variety of such contexts: One must be a competent physician in order to be board certified, a competent scholar to receive tenure in a university, etc. Competence establishes a level of ability for authorities to determine who is to enter the profession, enter the tenured ranks, and the like. Any concept that serves this function is inherently normative in the way it is used *to establish the abilities and level of abilities* — a normative choice of entry criteria — and to certify a person who possesses such abilities.

Although empirical investigation can establish whether a person has the abilities required for competence, the *choice* of criteria or tests of incompetence is not an empirical matter. Some evaluation is required to establish the abilities assumed as *the* requisite abilities in any context. For example, competence in driving, surfing, and nursing is established by well-entrenched contextual and evaluation criteria. The operative criteria might have been different than they are, and can shift over time. Thus, it is a mistake to infer that empirical judgments of psychological competence are free of prior evaluative commitments. The reverse is true: they are inescapably value-laden.

In its gatekeeping function, a judgment of competence may determine which persons should be treated as responsible choosers and therefore held responsible for their decisions, choices, and authorizations. A minor child whose decision to donate a kidney to a sibling meets all of the normal conditions and requirements of competence for adults — and indeed all of the conditions of informed consent — might not be considered competent because of age. By contrast, a minor child with venereal disease is often treated by statute as competent to consent to treatment without parental permission, regardless of his or her actual abilities. When a minor becomes a non-minor by turning age 18, more is at stake than becoming 'competent' to perform tasks such as voting and to engage in activities he or she was not previously 'competent' to engage in. We use age as a convenient, but not always accurate, threshold to indicate that sufficient knowledge, intelligence, experience, and sense of responsibility have been accumulated so that the person qualifies to pass through the gate. The non-minor thereby suddenly gains a string of rights that he or she is now deemed 'competent' to exercise, including rights of informed consent and refusal.

'Competent persons' and 'legitimate authorities' have a strong connection in this analysis, which in effect specifies that being competent is a sufficient condition of legitimate decisional authority, and therefore of the right to decide. However, there are exceptions to this rough gatekeeping rule. On some occasions, valid reasons may exist for overriding even the competent person's decisions, in which case another party becomes the legitimate authority. For example, for reasons of parental rights or public policy, we may allow parents to decide various matters for older teenagers who are competent choosers, or unions to decide health care rights and interventions for workers who are competent choosers (e.g., in matters pertaining to epidemiological research on asbestos exposure in factories). The reverse relation also may occur: we might allow an incompetent person to be the legitimate authority in some cases. For example, a retarded person might be permitted to choose his or her own proxy, or an eleven-year-old might be permitted to authorize treatment for venereal disease. Social mechanisms, such as requirements governing informed consent, cannot be strangled by wooden doctrines that render competent persons the *only* valid authorizers, come what may. We can tend the gate by saying that the competent person is presumptively the legitimate authority, and the incompetent person presumptively not the legitimate authority, but that compelling moral reasons, including other conflicting but legitimate exercises of authority, can alter the presumption. This problem, however, invokes concerns well beyond those in this paper.

C. *Partial Competence, Evanescent Competence, and Substantial Competence*

A person can be incompetent to perform some tasks x and y at the same time the person is competent to perform other tasks q and r. For example, a mildly retarded person correctly judged incompetent to handle *financial* affairs might nonetheless be competent to decide whether to participate in various forms of research. Blunt gatekeeping instruments tend to ignore such subtle differences, sometimes not so benignly.

More important is that some persons are incompetent to do something at one point in time and competent to perform that same task at another point in time. An individual's ability to consent, refuse, resist, or decide may thus fall under the designation 'intermittent competence'. Such matters can be complicated by the need to distinguish categories of illness that result in chronic charges of intellectual, language, or memory functioning from those that are characterized by rapid reversibility of these functions, as in the case of transient ischemic attack, transient global amnesia, and the like. In some of the latter cases, patients may be competent even from hour to hour. In an important decision on the rights of involuntarily committed mental patients to refuse treatment, one court encountered a highly intelligent patient who was sometimes competent to refuse medication, sometimes not, "depending on the day" [21].

The law has traditionally presumed that one who is incompetent to manage his or her estate is also incompetent to vote, make medical decisions, get married, and the like. Such laws are usually aimed at the protection of property rather than persons, and so are ill-suited to medical decision-making. Their global sweep, based on a total judgment of the person, can easily be carried too far, as courts have increasingly come to recognize. To say "Person X is incompetent to Y" should never carry the automatic implication that X is not competent to Z — i.e., not competent to perform an action other than Y. Such judgments defy much that we know about the etiology of various forms of incompetence — e.g., in the mentally retarded or in psychotic patients or patients with uncontrollably painful afflictions.

Competence should also be understood as resting on a continuum. Judgments of more or less competence reflect the fact that an ability can be possessed in different degrees. A person can also be more or less competent than he or she used to be, or more or less competent than another person. For legal and policy reasons, we specify various thresholds on the continuum that runs from wholly competent through partially competent to wholly incompetent. These cut-off points allow judgments that everyone falling on the side designated *competence* is, for policy purposes, equally competent, and everyone falling on the *incompetence* side incompetent [27].

But where do we draw the line? What is the proper criterion of 'sufficient' incompetence? How competent is competent enough? These problems are central for practical decisions about the competence to consent. We need to identify both the relevant abilities — the abilities necessary to give an informed consent — and the *degree* to which these abilities must be compromised in order to determine what counts as incompetence. Yet, it is rarely the case that individuals are equally competent on any *continuum* of competence, and it is an evaluative matter, as we shall later see in Parts II and III, how and where the threshold line marking incompetence is drawn.

D. *Competence To and Competence In*

We saw earlier that *general* competence requires a broad range of general abilities and that *specific* competence requires some set of specific abilities. Both are obviously ability-centered notions. However, we sometimes look not to abilities of the person but to the person's actions in order to see if a person is *acting* competently. Here we do not ask whether the person is competent *to* perform X, but whether he or she is competent *in* the performance of X. Ordinary language is not as precise as I shall here present it, but some rough distinctions can be drawn and should be drawn for practical as well as theoretical reasons. These distinctions can be schematically arrayed as follows:

I. General Competence ⎱
II. Specific Competence ⎰ ⟶ Abilities of the person
III. Competent Action ─────── Exercise of the person's abilities

In using the language of abilities of the person (or capacities, etc.), we should be as specific about the ability in question as possible, especially when discussing (II). For example, schizophrenics have various severe *psychological* problems, but it does not follow from this fact alone that they have *information processing* problems in particular. (Indeed, some studies show that schizophrenic patients comprehend reasonably well and are more aware of risks and benefits than the average nonmental patient ([25], [24].) In both (I) and (II), I shall speak of persons as *competent to* perform. In assessing actions, by contrast, I shall use the language of *performed competently* (or incompetently). The latter is what I mean by *competence in* — i.e., competence in the performance of some action. However, I shall primarily be discussing *competence to*, because of its ability-centered character.

Consider some specific abilities that reside in persons making them competent to perform some specific range of tasks — e.g., selling insurance, preaching, or playing the piano. A *person* who is competent to so such things may *act* incompetently in a specific situation, as when a competent insurance salesman or preacher fails to prepare for sales or sermon, and therefore fails at the assigned task. Patients who are competent deciders and therefore normally capable of decision-making tasks, may similarly fail. They may even fail miserably if ill in a hospital, overwhelmed by new information, ignorant, manipulated by presentations of data, etc. Despite performance of *an incompetent action*, the patient may still be a *competent person* or retain *a competence*. Even in circumstances of failure, a competent person is not stripped of the relevant abilities defining general competence or specific competences. A patient, then, may possess decision-making abilities and so be competent, but may not be 'able' to make a decision about treatment because not yet provided with the information needed to exercise decision-making abilities. Here we have a perfectly competent person who cannot competently decide (act) in the circumstances.

Consider an example remote from medicine: Mr. Zimmer is an avid and knowledgeable purchaser of electric utility stocks, with several years of purchasing experience. His knowledge of stocks, when to purchase them, when to sell them, and the like, is limited to utilities. Recently, he has become interested in gold and silver, but he knows little about gold and silver markets. His broker has strongly recommended silver, on the expectation of an imminent rise in prices. The broker explains the basic elements of precious metal markets, including the risks and potential rewards. Mr. Zimmer knows from his background in utilities transactions that *all* markets are tricky and that experts always second-guess any broker's recommendation. Although he has confidence in his broker, he knows the broker to be a man of precipitous and overenthusiastic judgment in volatile markets. On the theory I have propounded above, is Mr. Zimmer (a) a competent person?, (b) competent to purchase utilities stock?, or (c) competent to purchase the silver recommended by his broker?

The answer is that he is not (c) — i.e., he is not competent to decide about silver or any similar matter — although he is both (a) a competent, autonomous person and (b) competent to decide about utilities. Notice that we are not now asking whether he is competent to decide about his broker's merits. Mr. Zimmer's *particular* decision-making incompetence is different. It is (c): He cannot competently address this entire area of decision-making, unless he studies the matter carefully and gains experience. His would be an imprudent, uninformed choice. Yet he is a competent decision-maker *in general* — no less so than the expert in silver markets. Unless new constraints on the person are introduced, nothing renders him other than generally competent.

Sometimes a person of general competence will nonetheless act incompetently in a circumstance where he or she is called on to act in accordance with certain means and goals. Hospitals know this problem well. Consider a patient hospitalized with an acute disc problem, whose goal is to control her back pain. The patient decides to manage the problem by wearing a brace. This is a method she has used successfully in the past, and she feels strongly that she should return to this treatment modality. This approach conflicts, however, with the advice of her physician, who strongly advocates surgery. When the physician — an eminent man who alone in her city is competent to treat her — asks her to sign the surgery permit, she is *psychologically unable* to refuse. The patient's fear of an assertive and, in her view, powerful physician, exaggerated both by her illness and by her passive personality, makes it impossible for her to act as she desires. She is a person of *general* competence, is competent *to* consent or refuse in general, but not competent *in* choosing on this occasion.

For many purposes, including consent to research and treatment, a *competent person* is widely regarded in our society as one who is able to understand and communicate relevant information, to weigh risks and benefits, and to make a decision about acceptance or participation in the light of such knowledge and in light of his or her relatively stable values ([20], pp. 57—60). Such a person is able to behave purposively — that is, able to choose goals, able to choose appropriate means to goals, and able to act in accordance with the chosen goals and means. To be competent, the patient or subject need not actually perform these tasks, but their successful performance is one proof of competence. I am not suggesting that this is the *best* account of the criteria of the competent person, but only that, roughly speaking, it is the received cultural view at the present time.

II. COMPETENCE AND AUTONOMY

By what general yardstick is a person (judged to be) generally competent (I)? The characteristics just mentioned above — judgment, communication, stable valuing, and the like — indicate that competence is closely tied, at least in our culture, to autonomy. Law, medicine, and to some extent philosophy presume a context in which responsible, independent, and mature decision-making are associated with both competence and autonomy. Our view of children reflects this evaluative perspective, and so, I shall argue, does the motivation behind informed consent requirements.

A plausible hypothesis is the following: The *competent* person — at least in consent contexts and no doubt many others — *is* the *autonomous* person. If requirements of general competence serve the gate-keeping function previously attributed to them, autonomy in the person is the key criterion; when satisfied, the gate is unlocked. Criteria of general competence would thus be identical with criteria of autonomy, and any tests to ascertain general competence would examine whether a person is autonomous. In short, on this hypothesis, a person is generally competent to authorize or refuse to authorize an intervention if and only if the person is autonomous.

There is something promising about this proposal, but it is also fraught with conceptual and theoretical difficulties and can dangerously mislead, as we shall now see.

A. *The Connection Between Competence and Autonomy*

One problem is that competence is less closely related in meaning to autonomy than may at first seem apparent. Autonomy means *self-*

governance; competence means the *ability to perform a task*. A performance could be nonautonomous but competent, as when a person flawlessly performs a task because coerced by another person or while under hypnosis. The coercion does not render the person's act incompetent, although coercion does render the act nonvoluntary and thereby nonautonomous. The person may be psychologically competent to make a choice, and may in fact choose competently, but without choosing autonomously. One might also be incompetent to perform a task and yet autonomously undertake its performance. Clearly, then, autonomy and competence have distinct meanings, despite their common conflation in law and medicine, which tend to treat the terms as synonyms.

The trick to solving this problem is to avoid confusing the meaning of a concept with its criteria or with an application or special use it may have. The *meanings* of 'autonomy' and 'competence' are obviously quite different. But the *criteria* of general competence — here, the competent person — may be expressible exclusively in terms of the condition(s) of autonomy. That is, being an autonomous person may be *the* solely sufficient criterion that we use in our culture for general (but not necessarily specific) competence.

Again, this seems a promising analysis of the relationship between the two notions, but the connection is still too crude and imprecise as stated thus far, because both 'general competence' and 'the autonomous person' operate at an almost intolerably abstract level. Moreover, the properties that make up the autonomous person tend to be more stable and more independent of social evaluation than are the properties that constitute the competent person. Whatever may in addition be packed into the concept, the autonomous person is one who is capable of independent, intentional, informed, and reasoned judgments and actions. Any viable theory of the autonomous person must accept something like these properties as necessary conditions of autonomy, because of their centrality to the concept. No directly parallel claim can be made for the conditions of the competent person. The ability to make *independent* judgments, for example, might not be central to some socially influenced criteria of competence. In China, for example, this criterion seems nonessential. Competence is a concept that can range out and be applied to many different sets of properties in a person. As we have seen, it is inherently relative to a cultural or institutional context that fixes the needed properties. But autonomy has a dissimilar, or at least less broad, more determinate, and more stable set of conditions. The reason the two notions come together so nicely *in our culture* is that we have *made* them come together by establishing autonomy as the condition — i.e., the criterion — of general competence.

In informed consent contexts, as hypothesized above, general competence may be established solely through the criterion of the autonomous person, which at a minimum includes the possession of the ability to decide or choose autonomously. But the matter is more complicated when we come to *competent informed consent*. When we gatekeep in an informed consent context, we might look either to general competence (I) or to specific competence (II), the latter being the specific competence to make a certain kind of decision — e.g., whether to cease a particular drug therapy. These criteria of specific competence to consent not only differ from the criteria of general competence, but serve a different function as well. The qualities that make a person autonomous and therefore a competent person qualify the person as one *from whom consent is to be solicited*; the qualities that make a choice autonomous are those that qualify *the choice as an autonomous consent*. It is always an open question whether the competent person (an autonomous person) can in any given instance proffer an informed consent. The *capacity to make* autonomous choices in general (I) is distinct form the capacity to *make* such choices *in the circumstances* (II), and the presence of either capacity is no guarantee of its *exercise* in any single occasion (III). The person who qualifies to give an informed consent may not be able in the circumstances to give one, or may be able to do so but fail to do so.

In summary, 'autonomy' and 'competence' are not synonymous; nor does 'autonomous person' *mean* the same as 'competent person', although being autonomous is the criterion of being competent. When we search for the competence to consent, we look to the *criterion* of autonomy. An autonomous, competent person may proffer a nonautonomous, incompetent consent, because the criteria of general competence to consent or a specific competence are not the same as the criteria of an informed consent, which involves an action or competence *in* rather than competence *to*. Thus, one who is competent to choose may not 'competently' choose in the circumstances.

However, I recommend — based on the above analysis — that the *latter* use of the word 'competent' be avoided, so that 'competence to consent' and 'competent person' are the only terms used to refer to the gatekeeper function. It will often be difficult to determine whether what appears to be an informed consent in fact fulfills all the conditions and requirements of an informed consent. The validity of at least some particular authorizations that seem questionable in quality of autonomy is probably best determined purely by evaluation of the *general* competence to consent of the patient or subject. Here general competence to consent would become the criterion of the validity of a particular consent, and this of course is exactly what is often done when the concept of competence is used for gatekeeping purposes. In clinical and legal practice, many tests and standards of competence employed in various contexts — e.g., for evaluating the capacity to make a will or a contract, to stand trial, to care for one's personal affairs, or to refuse medically indicated treatment — apply less to the particular decisions in question than to the general abilities of the decision-maker.

It might be objected to this analysis that if *consent* is autonomous, it is irrelevant whether the *person* giving consent is autonomous. But this objection misses the most important connection between autonomy, competence, and informed consent, a connection already hinted at in much of the informed consent literature, where it is asserted that competence — along with disclosure, comprehension, voluntariness, and the like — is one of the *elements* of or *conditions* of informed consent, and indeed is essential to the meaning of 'informed consent' ([18]; [20], chaps. 3-4; [22], p. 2473; [12], p. 5). I do not believe that competence is best analyzed as a condition alongside other conditions or as an 'element' of informed consent. The connection between competence and informed consent is simply that competence judgments are the gatekeeping judgments in light of which persons are allowed or not allowed to make decisions for themselves. There is no other 'deeper' connection.

If this analysis is correct, then the characteristics of the person competent to give an informed consent *are* the characteristics of the autonomous person, but are not the criteria of an informed consent.

B. *But What is Autonomy?*

An ambiguity surrounds the concept of autonomy at work in my analysis. The meaning of 'competence' has now been discussed, but what about 'autonomy'?

Philosophical writings — and non-philosophical literature as well — link the concept of autonomy to an enormously broad spectrum of ideas. Analysis of the concept has been dominated in the last two centuries by Kantians, but this approach has now receded in significance and other viewpoints have proliferated as autonomy has enjoyed increasing popularity in philosophical theories. So diverse has the notion become that it can refer equally to duty, a right, a freedom, a disposition, or an action. The following have all been seriously proposed as philosophical explications of the central meaning of 'autonomy' (see [7], [4], [28], [16], [8]): "authenticity", "obedience to self-prescribed law", "obedience to moral law", "personal choice", "moral choice", "the freedom to choose", "having preferences about one's preferences", "choosing or creating one's own moral position", "mental health", "conscientiousness", "responsible action", and "accepting responsibility for one's views and actions".

In light of this semantic ambiguity, it is evident that autonomy is not a univocal concept in either ordinary English or contemporary moral philosophy. But if only various families of inconsistent ideas constitute 'the concept' of autonomy, then any unified and coherent theory of autonomy inevitably will refine or restructure the broader borders of the concept in light of the particular analytical objectives of the theory. Such a theory would have to be a revisionary or reconstructive analysis of the concept, not an ordinary language analysis.

Development of such a theory of autonomy is a complicated business, and one has yet to be produced, at least in comprehensive form. Unlike some of the more influential accounts of autonomy flourishing today — e.g., Kant's, Rawls's, Dworkin's — I incline toward a minimalist account of the properties one must possess to be autonomous. Predictably, this also inclines me to a minimalist account of the properties one must possess in order to be competent to choose. I am inclined, for example, not to require either *moral* conditions for autonomous choices (a theory of moral autonomy) or conditions of (second-level or authentic) reflective acceptance of one's motives and values, to cite two influential accounts. Instead, I would argue that a free person who can comprehend information and act intentionally qualifies as a person who can choose autonomously: The autonomous person must be able to grasp and appreciate the significance of pertinent information, form relevant intentions, and neither be controlled by internal forces that the person is not capable of resisting nor be controlled by external forces or influences that the person cannot resist — e.g., the coercive or exploitative interventions of others.

This view needs more detailed argument in its defense than can even be begun here. But it follows from my analysis that such a theory of autonomy is likely to be a decisive matter for a theory of the competent chooser. I say 'likely' because we have seen that there is no necessary connection between the two. Some non-minimalist writers on the theory of autonomy look to the autonomous person as one who is a free thinker, an innovator, a highly disciplined personality, or even a person of heroic independence. This characterization of the autonomous person as the exceptional person can have little to do, directly anyway, with the gatekeeping function served by the concept of competence.

III. EVALUATION IN THE SELECTION OF CRITERIA AND TESTS

A. *The Nature and Type of Incompetence Tests*

No authoritative legal, moral, or medical ruling specifies how incompetence to consent should be determined or tested for, but many tests have been proposed. In their valuable surveys of this literature, Loren Roth, Alan Meisel, and their colleagues at the University of Pittsburgh have analyzed this variety into seven cluster-categories of widely used 'tests', each of which specifies an operational criterion of incompetence ([22], [2]). The Pittsburgh Group, as I shall refer to them, has also attempted to show empirically that if some of the more frequently adopted tests are relied upon, large percentages of some populations of patient who voluntarily enter hospitals — e.g., psychiatric patients — are incompetent to consent to their own admission, let alone to treatment ([13], [14], [15], [22], [23], [24]).

In their analysis, tests for incompetence range from procedures to determine the mere absence of any indication that a person can make a choice, through various forms of the inability to understand or reason, to the inability to reach a reasonable decision. Some of their so-called 'tests' seem rather to be *criteria* of competence that set thresholds; but I shall look beyond this problem and assume that some means of testing can be devised to make any criterion *operational as a test*. As a rough generalization, a *test* is usually constructed to test for the *abilities* selected as well as for some *threshold* level of those abilities. The selection of a test follows from a prior choice of criteria of competence.

'Test', then, properly refers to procedures for determining a person's abilities. Among the tests the Pittsburgh Group has collected, some are more difficult to pass than others because some require either a higher level of skill at a defined task or an increased number of skills. In informed consent contexts, some of these tests could be and have been stringently employed, making it very difficult even for ordinary patients to qualify as competent to give a consent. For example, it might be required that a person be able to understand a therapy or research procedure without introducing a distorting bias, be able to weigh its risks and benefits, and be able to make a decision in the light of such knowledge even if the person chooses not to utilize the information.

The following schema (which involves some reshaping and reordering of the Pittsburgh Group's analysis) expresses the range of inabilities required in various tests for incompetence, roughly moving from the test requiring the least ability (the test the largest number of persons likely will pass as a *competence* test) to the other end of the spectrum of difficulty:

1. Inability to evidence a preference or choice.[1]
2. Inability to understand one's situation (or relevantly similar situations).[2]
3. Inability to understand disclosed information.[3]
4. Inability to give a reason.[4]
5. Inability to give a rational reason (although some supporting reason(s) may be given).[5]
6. Inability to give risk/benefit-related reasons (although some rational supporting reasons may be given).[6]
7. Inability to reach a reasonable decision (as judged, e.g., by a reasonable person standard).[7]

These tests cluster around three kinds of abilities or skills. Test #1 looks for the simple ability to evidence a preference. Tests #2 and #3 probe for abilities to understand information and to appreciate one's situation. Tests #4 through #7 relate to the ability to reason through a consequential life decision, although only Test #7 restricts the range of acceptable *outcomes* of a process of reasoning. These tests could be used either alone or in combination in order to determine incompetence.

Regrettably, our evidence for someone's lacking an ability that leads to a declaration of incompetence is often far less reliable and precise than we would like. In the extreme instance, we may even think a person lacks reasons when in fact the problem is that their reasons are unintelligible to us — i.e., not a part of what *we* recognize as an acceptable basis for human action. The continuum of our *evidence* for and beliefs about incompetence thus may vary dramatically from the actual reality of incompetence in the person.

B. *Moral and Policy Considerations in the Selection of a Test*

A decision about which test or tests to use in order to determine competence depends on a number of factors. Obviously, one's conception of the abilities required to give a meaningful consent is critical. More controversially, the choice of tests depends on moral and policy considerations having to do with the range of moral principles to be considered and balanced. In competence determinations, the tradeoff is generally between the moral principles of respect for autonomy, on the one hand, and beneficence, on the other hand. For example, if one is especially concerned about preventing abuses of autonomy, one may accept Test #1 as the only valid test of incompetence, or perhaps only Test #1, #2, or #3. On the other hand, if one's primary concern is that sick persons receive the most professional treatment possible, one may require patients to pass all these tests, or at least Test #6 or #7.

Those who accept a stringent test of incompetence (like #6 or #7) will place medical benefits and the safety and welfare of patients above their liberty or autonomy; a therapist accustomed to paternalism, for example, is sure to rely on a stringent test. Those committed broadly to liberty and autonomous rights, even for sick and psychiatric patients, will determine that these values take priority over medically-oriented values of health and welfare, and will therefore be attracted to a test that makes it more difficult to be found incompetent. There is no need to condemn either approach; respecting autonomy and protecting patients from substantial harms (or loss of health benefits) are both laudable goals. Too much deference to 'autonomy' can cause avoidable and undue harms, while disregarding expressed preferences may present a less significant risk to autonomy.

Conflicts based on these different moral commitments are the products of the most pervasive clash of principles in medical ethics, *viz.*, between the moral principles of respect for autonomy and beneficence ([5], Ch. 2). This need to balance respect for autonomy and beneficence also helps us see how competence and incompetence are examples of what W. B. Gallie calls "essentially contested concepts". An "essentially contested concept" involves dispute (the contrast) about criteria of its application that is a necessary (essential) feature because of the kind of concept it is. As Gallie puts it, "the proper use of [these concepts] involves endless disputes about their proper uses on the part of their users" ([9]). There are certain to be such competing conceptions of (1) the abilities that qualify to be criteria of competence, (2) the appropriate threshold levels of these abilities, and (3) tests of competence. The different policy options and divergent moral goals sketched in the earlier sections of this paper make such an outcome quite predictable.

The Pittsburgh Group has argued that "the search for a single test of competency is a search for a Holy Grail" that "will never end" until we face the fact that multiple conceptions of competence abound ([23], p. 283). It is easy to see why such confusion prevails, netting an essentially contested concept. Courts and legislatures are the meeting rooms of social values and multiple criss-crossing interests. Not surprisingly, the courts have failed to present a unified or developed doctrine of competence, tending instead to use quaint abstractions such as "having a sound mind", and leaving it to clinicians in particular cases to determine whether such vague standards are satisfied. Many different abilities can be encompassed in such broad general ideas, and several competing theories of comprehension, rationality, reasonableness, and freedom have been advanced to try to account for some of them ([23], p. 280). Disputes about incompetence often reflect a balancing of these weighty and sometimes competing considerations.

C. *Psychological Competence and Legal Competence*

Although the selection of tests and the criteria of competence and incompetence that underlie them is certain to be controversial, once the criteria for application of the concept of competence have been established — i.e., chosen from an evaluative point of view — it is then in principle an objective, empirical, and testable question whether someone *has* the abilities. That is, empirical inquiry can, in principle, be determinative whether the person passes the tests. One who satisfies the appropriate criteria of competence exhibits what is often referred to in the literature as *psychological competence*, under the assumption that the requisite abilities are psychological. Failure of the test warrants a judgment of *psychological incompetence*.

Here we must not confuse labels with states of mind. One could be labelled psychologically incompetent — by *the* 'definitive' test — when one was actually psychologically competent. As we saw in Part I, a test is simply a device for confirming, justifying, or rejecting the label and the sequence of events it invites; like any schoolroom test, it can fail to serve its intended objective. We can and should carefully distinguish *criteria* of competence from tests for it. The *conceptual* reasons set out in Part I and the *practical* reasons now under consideration both demand such care. Patients with problems such as 'locked in' syndrome or pure motor aphasia may be competent and may be able to communicate with others if only it can be discovered how to judge their behavioral responses. Such patients may be perfectly competent to consent, although we are not competent to communicate with them. To apply the label 'incompetence' to these patients may serve a useful function in the context, and may be necessary for legal reasons, but it is nonetheless mistaken as a judgment of psychological competence, whatever the outcome of an administered test.

Legal competence, by contrast to psychological competence, has to do with legal capacity (or incapacity), as a category distinct from psychological capacity. Some persons, such as precocious minors, may have psychological ability, but not legal 'capacity'. Some persons may have legal capacity without psychological capacity. Despite the contrast, however, legal competence generally builds on psychological competence, and adds an explicit, new evaluative dimension different from the evaluation involved in selecting abilities or tests of psychological incompetence. To say that someone is legally competent is to say that no one is justified in authorizing interventions in (some range of) the person's affairs or in acting on the person's behalf. The person's rights — including autonomy rights — cannot be preempted or overridden (for reasons pertaining to competence, anyway).

Legal incompetence is sometimes referred to as *de jure* incompetence, meaning that an individual is deprived of decision-making authority as a matter of law — whether he or she should be and whether he or she has the requisite psychological capacities. Declarations of legal incompetence, then, serve in various ways to assert that it is justifiable to authorize interventions, establish guardianship, and the like. Such declarations are usually applied to persons who are psychologically incompetent to conduct most of the common affairs of life. The law's special focus is less on psychological inabilities to function in society than on the consequent inability of persons to assume responsibility for their performance, which can be tied to such convenient criteria as age, linguistic facility, and knowledge of printed materials.

This concern is manifested by vesting in a guardian, tutor, or curator the power to assert the incompetent's rights. (Unlike the case of children, ownership of the person is not in question for guardianship, and thus the rights of the ward can only be exercised by a guardian within constraints of benefit.)

The law has long struggled to find ways of assessing whether a person is sufficiently in control psychologically to perform various societal functions. Legal competence often invokes concepts of rationality and reasonableness, applying various standards of the ability to comprehend information and to reason about the consequences of actions and decisions. Competence to give an informed consent thus can be decisively determined by legal standards of competence, and the abilities needed to give an informed consent may be identical to the range of abilities needed to be judged legally competent. But in the law, competence is not a unified concept with a single specified context. Notions and tests of legal competence arise from a variety of contexts such as property, estate, and business affairs, as well as from problems of individual responsibility for criminal behavior. Different forms and standards for determining competence thus have been delineated to fit the context in which a judgment is needed.[8]

Moreover, the law has never attempted to unify or standardize a concept of competence across its various interests. In keeping with its practical focus, the law's definitions and tests of competence have concentrated on area-specific conceptions. The guilty plea test, for example, is a relatively simple standard, whereas a fitness-for-trial test looks to a range of abilities necessary for meaningful participation. Both tests, and others in law as well, establish conditions under which an actor should be held to, or excused from, the consequences of actions. Still another set of standards of competence measures the substantive requirement of criminal responsibility, upon which hinges society's willingness to blame and punish actors for their behavior in addition to holding them financially or otherwise liable to face the consequences. Each such test of competence addresses both the specific context and the social importance of that context. Legal competence is therefore not merely a matter of value-neutral testing of psychological abilities to act.

D. *Medico-Legal Competence*

It is tempting to suppose that judgments of patient incompetence in medical contexts are analogous to psychological incompetence rather than legal incompetence, because psychological abilities are assessed by the physician, usually by psychiatric theories. However, this hypothesis neglects how medical or psychiatric determinations of incompetence often *function* like declarations of legal incompetence. In an institutional setting, such declarations serve to authorize second-party decisions or interventions, just as legal declarations do. Declarations of 'mental incapacity', for example, often function to authorize interventions or to recommend proxy decision-making.

Decisions whether to override a person's autonomy rights in order either to protect health and safety *or* to protect autonomy at a risk to health and safety often involve similar moral choices that again exhibit the balancing of respect for autonomy and beneficence [18]. Willard Gaylin has shown, convincingly I think, that our willingness to apply the word 'competent' to children of various ages will depend on whether we think we should grant liberty rights to the child or be cautious in extending such rights because the child might be harmed if we did. Gaylin suggests that the proper way to use the word 'competence' in regard to children (and others as well) is to be reluctant to confer it if an *important* value is at stake, and to be generous in conferring it when *little* is at stake ([10], pp. 35—38).

The Pittsburgh Group has noted that a criterion of any acceptable *test* of competence is that it be "set at a level capable of striking a balance between preserving individual autonomy and providing needed medical care" ([23], p. 280), and the President's Commission for the Study of Ethical Problems has similarly concluded that the "prudent course" in determining competence is "to take into account the potential consequences of the patient's decision". If the "consequences for well-being are substantial", the level of capacity required for competence should be increased; if the consequences are less significant, the level of capacity required may be appropriately decreased. "Thus a particular patient may be capable of deciding about a relatively inconsequential medication, but not about the amputation of a gangrenous limb" ([20], p. 60).

These reflections suggest — appropriately, in my view — that thresholds of competence should be graded and established in terms of risk (among other possible criteria). One reason for fixing the adjustable threshold of competence on this basis — although it is not a solely sufficient reason — has as much to do with the *consequences* of allowing a person to exercise abilities as with the level of *abilities* the person possesses. Just as requirements for obtaining informed consent should be increased as the risks and related possible consequences of the procedure increase (think of the elaborate consent form signed by Barney Clark), so the standards of competence do and should increase with the consequences for the affected party. Such movable thresholds could entail that different competence tests might be appropriately used for the *same* patient in the instance of *different* procedures. Also entailed is the acceptability of using various of the seven tests set out above (or something like them), as the level of risk increases. If these tests rest on levels of difficulty of requirements (or skill), then a level of difficulty in testing can be chosen that is commensurate with the level of risk involved.

From this analysis, it also follows that our moral judgments of beneficence and respect for autonomy are inescapably intertwined with our determinations of what will, in the circumstances, count as competence. More strongly worded, there is no such thing as competence apart from our *moral* judgments about where appropriate thresholds *should* be set. There is no threshold without a moral judgment, and no competence without a threshold.

What I have referred to as 'legal competence' is sometimes referred to in medical contexts as 'medico-legal competence'. This may be the reason that Albert Jonsen, Mark Siegler, and William Winslade dispense the following advice to practicing physicians in their handy pocketbook of clinical ethics:

Physicians often misunderstand the notion of competence ... [Competence judgments are] heavily value laden, resting more on norms than on facts. The term 'clinical competence,' used to evaluate physicians, carries this meaning ... *The terms 'competence' and 'incompetence' should be restricted to the legal status of a person: The person is judged, by proper legal authority, usually a judge, to be able to understand the nature and consequences of decisions* ... It should be clear that a judgment of competence or incompetence is the outcome of a specific legal process.

In the medical care situation, competence may have yet another meaning. Persons in need of medical care sometimes appear disoriented, confused, obtunded, psychotic ... For such situations we prefer the terms 'mental capacity' and 'incapacity' ... ([114], pp. 56–57).

This proposal, however, scarcely reflects the reality of the situation in contemporary medicine, where physicians' practices are often informal and at odds with legal standards. In one of their valuable empirical studies of incompetence and refusals by patients, the Pittsburgh Group discovered that "the staff in several of the hospitals studied appeared to have a uniform rule of thumb to determine incompetency, namely, that a patient be disoriented and incoherent — a stricter standard than most courts would use". Once the label 'incompetence' became attached, it was almost invariably felt by the staff that arm and leg restraints, Posey vests, anti-psychotic medications, and the like would follow routinely, and without solicitation of consent. These investigators note that

[T]he difference between these practices and those in many psychiatric hospitals, which also often deal with incompetent patients, is quite striking. As a result of a number of court decisions and statutes, psychiatric facilities in many jurisdictions are required to obtain formal determinations of incompetency and substituted consents from legally authorized proxies ([3], p. 474).

In summary, in this section I have discussed various types of evaluative judgments at work in judging competence: First, the necessary *abilities* must be fixed by an evaluation. Second, the precise *level of the abilities* must be fixed — a threshold determination. Third, some structure of *justified* authorization and intervention must be established. Although competence judgments require criteria that are evaluative, the legal system need not be involved. Psychological incompetence does not entail legal incompetence, and a person who is psychologically incompetent may retain legal authority to make personal decisions. Similarly, legal incompetence does not entail psychological incompetence; persons declared legally incompetent have on many occasions proved to be psychologically competent. Institutionalization entails neither form of incompetence. Indeed, institutionalized populations deserve special protections to assure that unwarranted inferences not be made merely because of the fact of institutionalization.

IV. CONCLUSIONS

Because competence functions as informed consent's gatekeeper the values underlying the choice of competence requirements play a pivotal role in informed consent. A strong beneficence orientation yields a requirement that can relieve physicians of the duty to obtain informed consent, but an autonomy-oriented approach would have the opposite outcome. Still, because informed consent requirements have as their goal the protection of autonomy, whereas competence requirements are intended to establish capacity for autonomous action in the person, there is no inherent contradiction in setting a difficult-to-meet competence requirement while thoroughly promoting autonomy through other informed consent requirements once competence is established.

It has commonly been assumed in literature on informed consent that a justified declaration of severely diminished competence or incompetence provides sufficient justification for an intervention on the patient's behalf without consent by the patient. Although such a rule has been implicitly followed by some physicians and judges, it does not follow from a person's incompetence to consent or refuse that the person's preferences may be justifiably overridden, or in any respect modified. Further, a person who is both legally and psychologically incompetent may still have preferences that should prevail (see [10], p. 38). We should disallow any logical connection between *incompetence* of any sort and *invalid* choice. Further, neither involuntary hospitalization nor coercive medical interventions from correct judgments of incompetence follow. The other side, of course, is that, while even fair and correct judgments of diminished competence or incompetence are not *solely sufficient* to justify recommended medical interventions or the nonacceptance of refusals by patients or subjects, a condition of substantial incompetence is generally *relevant* to the justifications of such interventions and nonacceptances.

Georgetown University
Washington, DC

NOTES

[1] This first test is simple: Can the person evidence a choice? The criterion is not the *quality* of decisionmaking, but whether a decision is made or can be made by the patient — e.g., by showing preference for treatment. If the patient cannot evidence a choice — say yes or no, shake his or her head, or engage in some other meaningful behavioral equivalent — then the patient is incompetent. Understanding and informed choice need not be in evidence. Unconsciousness, muteness, psychotic thought disorders, and marked schizophrenic ambivalence are standard examples of incompetence under this criterion, but there are numerous borderline cases.

[2] This criterion tests whether the patient or subject has the *ability* to understand, rather than testing the person's *actual* understanding. In particular, the person should be able to understand the nature of his or her situation and the nature and consequences of the

procedures in question. The determination of this ability can presumably be made inferentially through intelligence tests, mental status examinations, and the like. The ability to comprehend and appreciate risks, benefits, and alternatives to treatment is emphasized. Precise or perfect understanding is not required; only fundamental cognitive abilities must be demonstrated.

³ The third criterion tests whether the patient is able to understand the meaning of the information provided in the professional's disclosure. Often this is determined by whether the patient did in fact understand, rather than whether the person is generally able to understand. Neither here nor in the previous test (#2) is the person's weighing of the information to be evaluated.

⁴ Failure to articulate a reason in support of the decision refers to choosing without being able to give any reason for the choice. Suppose a patient refuses to have surgery, and is asked why. If she is unresponsive, staring blankly at the wall, or simply says "I don't know", she is judged incompetent under this criterion, despite her explicit refusal.

⁵ In the fifth test, the patient is able to give a reason, but not a rational reason. How 'rational' should be defined has attracted sharp argument, including argument about whether many of our most ordinary acts are based on a rational reason. Usually a minimal notion of rationality is used. For example, there must be some plausible connection between the reason and the choice. Quality of thought is presumably the determinant, although no one has been able to delineate a theory that satisfactorily distinguishes rational and nonrational reason, and disagreements abound over problems of subjective assessments of rationality, the nature of mental illness, and the like. Sometimes an 'insane delusion' criterion is invoked in law, for example, in voiding a contract or invalidating a will, in order to hold that a person is not a freer and responsible agent.

⁶ This test requires that a subject fails to base a decision on a reasonable weighing of potential risks and potential benefits (perhaps using a Baysean or some other expectancy-value model of choice). The person could have some rational reasons yet fail to employ risk/benefit comparisons. Thus, this test is stiffer than the previous five. It also runs the risk of being the most arbitrary, because it requires a weighing of items labelled by others as 'risks' and 'benefits' — whether the patient sees them as risks and benefits or not.

⁷ This test bases judgments of incompetence on the outcome chosen rather than on the process by which the decision was reached. Usually the appeal is to an 'objective' criterion such as whether the patient failed to choose what a reasonable person would choose in similar circumstances. A time-honored example of this test is the determination whether a person is able to take care of himself or herself, especially to manage property or maintain health, on the basis of his or her actual decisions. An outcome might also be invalidated in an informed consent context if no reasonable practitioner would sanction the choice as medically sound. This test is the one most likely to conflict with autonomous choice.

⁸ The following four contexts and their connections to competence have been noted by Donald Bersoff:

(1) *Competency to stand trial*: Defendants must be capable of understanding the nature and purpose of the proceedings against them. They must be able to comprehend their own status and condition in reference to the proceedings. They must be

capable of assisting the attorney in conducting their defense or be able to conduct their own defense in a rational manner.
(2) *Competence to plead guilty*: Those accused must not be so mentally impaired as substantially to impair ability to make a reasoned choice among the alternatives presented and to understand the nature of the consequences of their plea.
(3) *Capacity to contract*: Those who make contracts must be able to appreciate and understand the nature and effect of the act in which they are engaged and the business they are transacting and to exercise their will in relation thereto.
(4) *Capacity to make a will*: Testators must have at the time they make a will sufficient ability to understand the business upon which they are engaged, the effect of the act of making a will, the nature and extent of their property, the persons dependent upon their bounty, and the mode of distribution among them. They must also have memory sufficient to collect in their mind the elements of the business to be transacted and hold them long enough to form a reasonable judgment.

This schema is taken from a longer, unpublished draft by Bersoff that was later shortened and published (anonymously) by the National Commission for the Protection of Human Subjects of Biomedical and Behavioral Research, in its *Appendix to Report and Recommendations: Research Involving Those Institutionalized as Mentally Infirm* ([19], chap. 6, p. 68).

Courts have disagreed on which of the following three abilities is most crucial to a determination of competency: (1) the ability to reach a decision based on good — or rational — reasons, (2) the ability to reach a *reasonable result* in a decision, or (3) the ability to *make a decision* at all. Each of these diverse criteria is an attempt to express the idea that the incompetent person is 'devoid of reason'. In case (1), courts attempt to determine whether persons have the abilities — to take some examples relevant to informed consent — to understand the nature of a procedure, to assess risks against benefits, and to come to a decision as a result. In landing on these particular abilities, the courts are here choosing among a long list of possible abilities.

The concept of 'rationality' is often invoked in unanalyzed and even provocative ways in the law, but it essentially functions as a principle of respect for autonomy. To say that an agent is rational is to say that he or she possesses a quality of autonomy that deserves respect for his or her decisions. This is usually judged in concrete ways by whether a person is choosing reasonably and sensibly.

Following (2), different courts look to abilities to make a decision that expresses a reasonable result. This criterion follows the logic of the concept of a reasonable person: Is the decision reached one which a reasonable person might make? If persons make self-harmful choices — whether as a result of illness, retardation, emotional weakness, brain dysfunction, or whatever — courts adopting this second standard find the person incompetent because these mental conditions lead to judgments a reasonable person would not make.

By contrast, courts following (3) look exclusively to the ability to make a decision, whether it be rational or reasonable. This standard involves less of a focus on specification of requisite mental abilities, yet it presumably would not regard as valid the consent of persons out of touch with reality. As in the case of (1) above, courts look to whether the person has an adequate understanding of such matters as the nature of the procedures, risks and benefits, and possible alternatives. If the person possesses the ability to grasp this information, then any decision reached would be a valid decision — even if not one a reasonable person would make and even if not closely tied to a 'rational' reason.

[9] I am indebted to Ruth Faden, Laurence McCullough, James Childress, Alan Meisel, Nancy King, and Bettina Schoene-Seifert for numerous helpful discussions on various of the arguments and distinctions in this paper. I owe a great deal of the substance and perspective in the paper to their criticisms and constructive proposals, running over a period of several years.

BIBLIOGRAPHY

1. Appelbaum, P. S., Mirkin, S. A., and Bateman, A. L.: 1981, 'Empirical Assessment of Competency to Consent to Psychiatric Hospitalization', *American Journal of Psychiatry* **139**, 1170—1175.
2. Appelbaum, P. S. and Roth, L.: 1982, 'Competency to Consent to Research: A Psychiatric Overview', *Archives of General Psychiatry* **39**, 951—958.
3. Appelbaum, P. S. and Roth, L.: 1982, 'Treatment Refusal in Medical Hospitals', in President's Commission for the Study of Ethical Problems in Medicine and Biomedical and Behavioral Research, *Making Health Care Decisions*, U.S. Government Printing Office, Washington, D.C., Vol. 2, pp. 411—477.
4. Benn, S. I.: 1976, 'Freedom, Autonomy, and the Concept of a Person', *Proceedings of the Aristotelian Society*, **113**, 124—128.
5. Beauchamp, T. L. and McCullough, L.: 1984, *Medical Ethics: The Moral Responsibilities of Physicians*, Prentice-Hall, Englewood Cliffs, New Jersey.
6. Culver, C. and Gert, B: 1982: *Philosophy in Medicine: Conceptual and Ethical Issues in Medicine and Psychiatry*, Oxford University Press, New York.
7. Dworkin, G.: 1978, 'Moral Autonomy', in H. T. Engelhardt, Jr., and D. Callahan (eds.), *Morals, Science, and Sociality*, The Hastings Center, Hastings-on-Hudson, New York, pp. 156—171.
8. Edwards, R. B.: 1981, 'Mental Health as Rational Autonomy', *Journal of Medicine and Philosophy* **6**, 309—322.
9. Gallie, W. B.: 1956, 'Essentially Contested Concepts', *Aristotelian Society Proceedings* **56**.
10. Gaylin, W.: 1982, 'The Competence of Children: No Longer All or None', *Hastings Center Report* **12**, 33—38.
11. *In the Matter of Guardianship of Richard Roe III*, 421 N.E. 2d 40 (Mass. 421, Supreme Judicial Court of Massachusetts, 1981).
12. Jonsen, A. R., Siegler, M. and Winslade, W. J.: 1982, *Clinical Ethics*, MacMillan, New York.
13. Lidz, C. W. *et al.*: 1984, *Informed Consent: A Study of Decisionmaking in Psychiatry*, The Guilford Press, New York.
14. Meisel, A.: 1981, 'What It Would Mean to be Competent Enough to Consent to or Refuse Participation in Research: A Legal Overview', in N. Reatig (ed.), *Proceedings of the Workshop on Empirical Research on Informed Consent with Subjects of Uncertain Competence*, National Institutes of Mental Health, Rockville, Maryland.

15. Meisel, A., Roth, A. H. and Lidz, C. W.: 1977, 'Toward a Model of the Legal Doctrine of Informed Consent', *American Journal of Psychiatry* **234**, 285—289.
16. Miller, B.: 1981, 'Autonomy and the Refusal of Life-Saving Treatment', *Hastings Center Report* **11**, 22—28.
17. Murphy, J. G.: 1979, *Retribution, Justice, and Therapy: Essays in the Philosophy of Law*, D. Reidel Publishing Company, Dordrecht, Holland.
18. National Commission for the Protection of Human Subjects of Biomedical and Behavioral Research: 1978, *The Belmont Report: Ethical Guidelines for the Protection of Human Subjects of Research*, DHEW Publication No. (OS) 78-0012, U.S. Government Printing Office, Washington, D.C.
19. National Commission for the Protection of Human Subjects of Biomedical and Behavioral Research: 1978, *Research Involving Those Institutionalized as Mentally Infirm: Appendix to Report and Recommendations*, DHEW Publication No. (OS) 78-0007, U.S. Government Printing Office, Washington, D.C.
20. President's Commission for the Study of Ethical Problems in Medicine and Biomedical and Behavioral Research: 1982, *Making Health Care Decisions*, Vol. 1, U.S. Government Printing Office, Washington, D.C.
21. *Rennie v. Klein*, 462 F. Supp. 1131, (D.N.J. 1978).
22. Roth, L. and Meisel, A.: 1981, 'What We Do and Do Not Know About Informed Consent', *Journal of American Medical Association* **246**, 2473—2477.
23. Roth, L., Meisel, A. and Lidz, C. W.: 1977, 'Tests of Competency to Consent to Treatment', *American Journal of Psychiatry* **134**, 279—284.
24. Roth, L. *et al.*: 1980, 'Competency to Consent to and Refuse ECT: Some Empirical Data', unpublished paper.
25. Soskis, D. A.: 1978, 'Schizophrenic and Medical Inpatients as Informed Drug Consumers', *Archives of General Psychiatry* **35**, 645—647.
26. Thompson, W. C.: 1982, 'Psychological Issues in Informed Consent', in President's Commission for the Study of Ethical Problems in Medicine and Biomedical and Behavioral Research, *Making Health Care Decisions*, Vol. 3, U.S. Government Printing Office, Washington, D.C., pp. 83—115.
27. Wikler, D.: 1979, 'Paternalism and the Mildly Retarded', *Philosophy and Public Affairs* **8**, 377—421, reissued in 1982 as 'The Bright Man's Burden', in R. Macklin and E. Gaylin (eds.), *Mental Retardation and Sterilization*, Plenum Press, New York, Chapter 10, pp. 149—166.
28. Young, R.: 1980, 'Autonomy and Socialization', *Mind* **50**, 565—576.

EDMUND L. PINCOFFS

JUDGMENTS OF INCOMPETENCE AND THEIR MORAL PRESUPPOSITIONS

As I read through Dr. Knight's paper [3], wondering what useful thing I could possibly say to a person so experienced in the day-to-day decisions concerning competence, given my own near-complete inexperience, it strikes me that this inexperience itself may have some small advantage. At least I can approach the problems afresh, innocent of partisanship toward any of the positions that have already been marked concerning the way to approach the difficult decisions that must be made by physicians and by patients. I will not be a paternalist, nor an advocate of unfettered autonomy for the patient, nor do I have an initial bias toward any of the many positions between. My contribution will be one that philosophers out of their ignorance (celebrated since Socrates) are privileged to make: the raising of questions, but most especially of questions about questions that others are asking.

I. WHEN IS A PATIENT COMPETENT?

What, exactly, are we looking for when we ask that question? What is it to say that a person is competent to make a decision concerning his treatment? Is it to say that he possesses some desirable but hard-to-define power, competency? Are we looking for the necessary and sufficient conditions for the presence of that power? Is competency a part of the description of a person in the way that his height and weight are a part of his description? If so, then, by contrast with these characteristics, competency seems elusive and obscure.

Competency may, in this respect, be conceptually akin to *mens rea* in criminal law. The guilt of a person who has done a legally forbidden act typically turns on the presence or absence of *mens rea*; and *mens rea* can, like competence, be thought of as a particularly elusive and hard-to-pin-down descriptive quality. To think of *mens rea* in this way is to be led to speak obscurely about 'the will', 'mental elements', and freedom. We amble down a path along which we find that we have traded difficult questions about one concept for equally difficult questions about a set of others. Similarly, we may be led, in pursuit of the necessary and sufficient conditions of competence, into the tangled thickets of the metaphysics of mind.

If this analogy between the conceptual problems of the philosophy of law and of medical philosophy is a reasonably close one, it suggests an approach to competence that may be helpful. H. L. A. Hart, in a famous essay [1], suggests that we think of *mens rea* not as a part of the description of the agent, but rather as a defeatable ('defeasible') claim concerning the condition of the accused. The claim, or judgment, expresses, according to his notion, a rebuttable presumption concerning the condition of a person who has committed a forbidden act. Thus, whoever kills another person is guilty unless he was insane, coerced, acted in excusable ignorance of fact, etc. These recognized heads of exception can defeat the verdict of guilty. To give an example as old as Aristotle, we would not hold a person guilty of murder who, in trying to save a man by giving him something to drink, in fact kills him (*Ethica Nicomachea*, III, I, 1111a 1—15). To say, then, that there was *mens rea* is shorthand for saying that none of the heads of exception applies in the presence of the forbidden act.

There may be merit in thinking of competence, too, as a defeasible concept. Given that the patient has expressed a preference for or against the prescribed treatment, we may defeat the presumption that he is competent only by application of one of the recognized heads of exception. To say that he is competent is simply to say that there is no sufficient reason for saying that he is incompetent. This approach leads us not into the mistier regions of metaphysics but into the examination of the sorts of considerations that can defeat the presumption of competence.

Given so much, we might now make use of some distinctions that have been found useful in the philosophy of law. We can begin by distinguishing questions of *eligibility* from questions of *encumbrance* [2]. To say that a person is eligible to be competent is to say that he is not an infant, not severely retarded, not in a coma, not relevantly insane. Here the patient is incapable of making any decision at all, either absolutely, in the sense that he is because of his relatively permanent condition, unable even to express a decision, or relatively, in that what he expresses cannot be assumed to result from deliberation. Deliberation requires the exercise of capacities that he just does not have. He may not have them permanently, or the capacities may temporarily not be there, as in delirium, shock, or confusion.

Ineligibility is, then, a characteristic of the patient. We will find it useful to speak of encumbrance as a characteristic of an eligible patient's decision. To say that his decision is encumbered is to say that it rests on ignorance of relevant facts, that it is made under perceived threat, the the patient is paralyzed by fear, etc. An unencumbered decision 'truly represents' what the patient wants done about the treatment in question. His decision is taken as representing him in a way analogous to the way the unencumbered accused's act represents what the accused 'really wanted to do'. But an unencumbered decision may still be a bad decision.

The question whether the patient has decided *well* should be distinguished from the question whether the decision was truly his, represents him, is unencumbered. This distinction is, I suspect, from some of the things Dr. Knight says, a source of trouble and uncertainty. He mentions that among the criteria suggested for incompetency is the very decision made by the patient if that decision is 'irrational'. The surgeons are not willing to accept the psychiatric finding of competency when a patient refuses to allow the radical neck surgery that will probably, with radiation, rid him of a growing tumor of the soft palate and tonsillar area of the mouth.

II. WHAT OUGHT TO BE DONE WHEN AN APPARENTLY COMPETENT PATIENT'S DECISION IS 'IRRATIONAL'?

'Irrational' is, of course, a weasel-word in philosophy. Beyond clear cases of self-contradiction, of failure to take account of evident and relevant facts, or of egregious failures to understand means-end relations, we are quickly led into a twilight zone in which one person's irrationality is another's rationality.

I am particularly struck by the cases cited by Dr. Knight in which it is taken for granted by some physicians that a death wish on the part of a terminally ill patient is regarded as evidence of irrationality, or in which the refusal of the above-mentioned patient to have his tumor removed is evidence of irrationality. The terminally ill patient may have reason to think himself, and those around him, better off if he were dead. The tumor patient may have reason not to want to live with a radically mutilated neck. We might not accept these reasons, but their being inconsistent with our notions of the way people ought to think is hardly ground enough for classifying them as irrational. They are, in any case, not irrational in the strong sense of that term in which inconsistency, failure to take account of relevant facts or of means-ends relations are irrational. There is not a clear case, as in the strong sense, of a mistake in reasoning.

To give weight to the distinction between irrationality in the strong sense and in the weak sense in which the decision seems just foolish or mistaken is to distinguish two different sorts of problem: the problem that arises when an unencumbered decision is demonstrably badly reasoned, and the problem that arises when the reasoning is in order but the conclusions still seem foolish or irresponsible.

Suppose we can show that the patient's conclusion does not follow, or that he is contradicting himself. Must we then allow his decision to govern what treatment he shall receive? Or whether he shall receive any treatment at all? The patient, by this hypothesis, has no logical right to his conclusion in which, say, he refuses treatment. *A fortiori*, he will have no logical right to his conclusion if he is unable to give any reasons for it at all. (This is not to be confused with the case in which the patient is reluctant to give his reason — as the tumor patient might be reluctant out of embarrassment that he should then be revealed as preferring death to disfigurement.) What, then, follows about the right of the surgeons to operate anyway? By hypothesis, the patient's decision is unencumbered, but, also by hypothesis, his reasoning is askew in that it does not carry him to the conclusion on which he now insists. Can it then be said that he does not know what he wants, that what he now says does not qualify as his decision? Would he decide differently if he could be made to see the flaws in his reasoning? How can this be answered *a priori* and without further specification of the facts of the case? How are we to know when reasoning is mere rationalization, so that the conclusion is what matters, not the reflective process that leads there?

Some people just do not reason well, or often do not reason at all. What are we then to say about the obligation of abiding by their ill-reasoned conclusions concerning treatment that they themselves may receive? Should we regard this inability or unwillingness to reason as another kind of encumbrance on the decision, or should we think of it as a failing that must be allowed to an autonomous agent?

But suppose that the unencumbered patient's reasoning is generally in order, but that his decision still seems foolish or irresponsible? He argues cogently to a conclusion the surgeons are unwilling to accept:

that he should not allow the neck surgery, for example. It is here that the term autonomy will almost inevitably enter the discussion; and it is to the role and force of that term in decision that we must, inevitably, turn.

III. WITHIN WHAT LIMITS CAN THE PATIENT'S 'AUTONOMY' BE RESPECTED?

Must the unencumbered, cogently reasoned decision of an eligible patient be accepted, on the principle that he is an 'autonomous' agent, free to decide those matters that most intimately affect himself? Here it is useful, I think, to distinguish foolish from irresponsible decisions, rather than to speak, ambiguously, of 'irrational' decisions.

To speak of foolishness in a decision is, typically, to speak of imprudence. What the patient decides is not in his own long-term interest, let us say. The tumor patient, it might be argued, should be willing to accept advice to the effect that effective and honorable lives can be and are led by badly mutilated persons, and that appearance should not weigh more heavily in his calculations than the continuation of his life. To speak of irresponsibility in a decision is, on the other hand, to speak the language of morality. What the patient decides is morally indefensible. The tumor patient, let us say, has no right to refuse treatment now, when it is certain that he will require far more difficult treatment later, treatment that will require facilities and services that could better be devoted to the timely treatment of other patients.

At least we can, at this stage of our analysis, put behind us what certainly seems to be a confusion in the surgeon's thought. Competence is one matter; foolishness and irresponsibility are another. It is illegitimate, as Dr. Knight implies, to reason from a decision that seems to you wrong, to the incompetence of the agent reaching that decision. That is to beg the question at issue, which is precisely what we are to say about the competence of a person whose decision it is. Thus, the President's Commission for the Study of Ethical Problems in Medicine and Biomedical and Behavioral Research wisely refuses to accept what they call the 'outcome approach' to the determination of the patient's competence to consent.

The outcome approach — which the Commission expressly rejects — bases a determination of incapacity primarily on the content of the patient's decision. Under this standard, a patient who makes a health care decision that reflects values not widely held or that rejects conventional wisdom about proper health care is found to be incapacitated ([4], p. 170).

Back to foolishness and irresponsibility, and to 'autonomy'. What is it, in these terms, that must be protected, in respecting the 'autonomy' of the patient? It may seem obvious that he ought to be protected in his right to make decisions that are foolish, but not protected in his right to make irresponsible decisions. He is to be allowed to be the judge of his own interests, all else being equal. But all else is not equal when the justifiably protected interests of others are adversely affected by his decision. What is obviously right in these global terms, however, would need a great deal of qualification on closer analysis. In particular, much would have to be said, and no doubt has been said, about the relations that obtain between the responsibility of the physician for the patient and the patient's responsibility for himself. The institutional context is crucial. While there may be a problem if our tumor patient has entrusted himself to the care of a surgeon in a hospital, the problem is either not the same, or does not exist at all, if the patient is walking the streets after a diagnosis of malignant tumor. The act of entrusting himself arguably alters the balance of responsibility for his health between himself and the physician. What is to count as such an act, then, turns out to be an important matter, but a matter that others here are better able to examine than I am.

Suppose, then, that (I do not analyze the term) the patient has 'entrusted' himself, and at the same time insists on a decision that is irresponsible. By hypothesis, he is competent to withhold consent, and he does so. What recourse, if any, is open to the physician? That is, of course, a question of institutionally defined procedure, but it is also a question what should be taken into account, and with what weight, in the operation of that procedure. The present question is what weight should be given an irresponsible refusal. It seems that, given even an irresponsible refusal of treatment, there is a presumption against treatment. To rebut that presumption, one would have to offer an analysis of 'entrustment' that makes it an irrevocable act; and it is doubtful that any prospective patient understands the act of entrusting his care to a physician as an irrevocable one. To put the point in terms familiar in contemporary Rawlsian political philosophy [5], it is doubtful that anyone would assent to a practice according to which 'entrustment' had that irrevocable consequence, if he did not know what role, patient or physician, he was to play in that practice. Given that the doctor-patient relationship is one that is ideally and usually freely entered into, then it would seem unlikely that the articles of agreement should specify that, should I play the role of patient, I could not refuse life-altering treatment, if, at the time, I were fully competent to make a decision.

To recapitulate, the analysis so far suggests that the patient's rejection of treatment (neck surgery in Dr. Knight's example) governs if three conditions are satisfied:

(1) The patient is eligible to make a decision: not an infant, not relevantly insane, not severely retarded, not in a coma, etc. In these cases, either (as in infancy or severe retardation) there is no prospect of the patient's thinking to some purpose in the future about what he wants done or not done; or (as in unconsciousness, temporary insanity, or confusion) there is such a prospect. The implications for paternalism of this distinction are reasonably obvious.

(2) The patient's decision is unencumbered: he is not ignorant, nor lacking appreciation, of relevant facts concerning the disease and its probable outcome if untreated, or concerning treatment and its probable outcome; neither is he under any direct or indirect compulsion or duress; nor is he paralyzed by fear, etc.

(3) The patient's reasoning is in order: he gives reasons, is not contradicting himself, is not making means-ends mistakes in reasoning, etc.

Given these conditions, the patient's decision to reject neck surgery must, I suggest, presumptively be respected. That presumption may be a defeasible one, but what is to defeat it? Not, I have argued, merely that he arrives at an 'irrational' conclusion. Given that the three conditions have been met, we can not argue that the patient's decision not to accept treatment is a sufficient condition of his incompetence. It does not seem that the presumption can be defeated by showing that the irrationality consists in the foolishness of the patient's failing properly to take into account his own presumed interests: the avoidance of extreme and extended pain and miserableness, in this case. But does the irresponsibility of the patient's decision entitle the surgeons to operate? I have denied this above, but the point needs closer examination.

IV. WHAT IS THE PHYSICIAN'S RESPONSIBILITY IF THE PATIENT'S DECISION IS IRRESPONSIBLE

The patient's decision to reject treatment, in our example, will probably result in difficult, time-consuming, and costly procedures later that will deprive others of needed medical attention. This is a consideration to which he ought, morally, give weight. He is also not giving proper weight to the consideration that the distress to and burden on his family will be greater in the end if he does not receive treatment now. What is the moral implication for the physician of this moral failure of the patient to give appropriate weight to the moral aspects of his decision?

The physician's position is, to say the least, not easy. He may, on the one hand, be a party to the non-formal and vague understanding that I have called entrustment. What the bounds on that understanding are is not clear from the beginning, and it appears that these bounds may change as the implications of entrustment become clearer to patient and physician. On the other hand, the physician is, willy nilly, the representative of the system of health care, and has a responsibility to see that it is not abused by unreasonable demands — as by hypothesis it will be in the case before us, if treatment is not given now. (I speak for simplicity of the physician's decision even though I realize that there may be a fairly elaborately structured corporate decision by a committee.)

I am inclined to think that the patient's right to withdraw is what should govern. I have suggested that it seems unlikely that you or I should enter into an irrevocable entrustment agreement if we did not know whether we were to play the role of patient or doctor. There must, I think, be the option to withdraw, although no doubt qualifications must be placed on that option. It cannot be exercised, say, in the midst of a non-anesthetized operation, nor at a crucial stage in a potentially dangerous procedure, such as an arteriogram. But there is also a point about integrity, a point that goes beyond the revocability or non-revocability of the entrustment agreement.

The point is that no one can take on himself the responsibility for another person's moral decision, nor can anyone give away his responsibility for a decision by handing it over to someone else. By hypothesis, we are speaking of a patient who is fully competent; we cannot set aside his decision on the ground that it is an incompetent one. It is irresponsible, though, and, by hypothesis, that is the only ground for overriding it. But also, by hypothesis, it is *his* decision, and hence his irresponsibility that is in question. By right of withdrawal from the entrustment agreement, he can decide, irresponsibly, not to receive treatment. If the physician, refusing to accept withdrawal, were to substitute his decision for the patient's, he would be violating what ought to be a part of the original agreement, a part that until now has not been obvious. For how, morally, can I enter into an entrustment agreement under which I authorize you to make decisions that I am morally responsible to make? I would be sacrificing my integrity as a moral person by entering into any such agreement. But I am morally responsible to make decisions concerning my own life or death, and concerning the alternative kinds of life that are open to me, given this or that medical procedure. The point is an Aristotelean one, to the effect that I cannot excuse myself from failing to do the things I ought in life, on the ground that I am incapable of doing them, if my incapability is the result of my failure to keep myself fit to do my duty. So I have a secondary obligation to keep myself fit. I cannot therefore entrust myself to a physician in such a way that it is his decision what risks I shall take; and I cannot leave it to him to decide under what post-operative disabilities, given options, I am to suffer.

None of this is to imply that the physician has a duty just to stand by and allow the patient, by his decision, to place an unjustified burden on the health care system. He can and ought to try to make clear the future consequences for that system of the patient's refusal to receive treatment now.

Thus, the duty to inform the patient extends, on my view, beyond the duty to make the patient fully aware of the possible or probable consequences for himself of accepting or rejecting the prescribed treatment. (This is the duty to help the patient avoid an imprudent decision.) It may also extend, as in the tumor case, to making the patient aware of the facts that bear on the irresponsibility of the decision. In the case before us, these facts would include not only the unwarranted future burden on the health care system, but the probable effects of an untreated tumor on the lives of family and friends.

V. CAN INCOMPETENCE BEAR THE WEIGHT THAT IS PUT UPON IT?

In closing, I should mention that it strikes me, as I read through Dr.

Knight's informative paper, and through the volumes published by the President's Commission, that there is a tendency to lay down the criteria of competency in such a way as to make possible the ruling out, as incompetently made, a decision that the physician simply regards as wrong. There seem to be a good many (at least four!) aces in the hole, in the 'game' played between physician and patient. We learn, for example, (1) that there is a possibility or probability that a patient facing major surgery is incompetent ([3], p. 3); (2) that when the patient does not want to 'cooperate with the prescribed treatment regimen', the patient's competence comes in question ([3], pp. 4, 21); (3) that there is *prima facie* incompetence, from the physician's point-of-view, if the patient wishes to die ([3], pp. 4, 13); and (4) that if the risk in treatment is minimal for a serious illness, and the benefit great, the patient who refuses treatment will be declared incompetent ([3], p. 14).

With respect to the first ace, for example, the presumption that a patient facing major surgery is incompetent, I can see no reason other than building a road-block against an 'irrational' patient-refusal for accepting this presumption. But if we are speaking, as we should be for clarity, of the unencumbered and cogent decision of an eligible patient, then presumptive competency has not been defeated whatever the patient decides. Similarly for the patient who has decided that he wants to die, or that he does not want to accept a minor risk for a great benefit. And to employ a low threshold test of competency for a patient who refuses treatment and a high one for a patient who accepts it is to stretch the cloak of competency to fit the 'right' decision and to shrink it so that it will not fit the 'wrong' one.

I should also mention that the insistence in Dr. Knight's paper that, beyond questions of general competency, there should be care that the competency tested for is the competency to make the particular decision at hand, seems to me to be proper, but also to contain a danger. It is proper because a person's general deficiencies may not be relevant to his capacity to make the decision at hand. But it contains the danger that the narrow focus on the capacity to make the decision in the particular circumstances of that decision, throws intense light on the seeming irrationality of the decision of a patient who, on general grounds, and with respect to his reasoning on analogous matters, is quite clearly competent. In short, the outcome criterion can quite easily slip in again through the back door.

If I have, in these remarks, succeeded in raising a question or two about the questions under discussion, I will, I hope, have justified the presence of this review here among people who know a great deal more than I do about the day-to-day problems of determining competence.

University of Texas
Austin, Texas

BIBLIOGRAPHY

1. Hart, H. L. A.: 1949, 'Ascription of Responsibility and Rights', *Proceedings of the Aristotelian Society*, New Series (XLIX), 171—194.
2. Hodson, J.: 1983, *The Ethics of Legal Coercion*, D. Reidel Publishing Company, Dordrecht, Holland.
3. Knight, J. A.: 1991, 'Judging Competence: When the Psychiatrist Need, and Need Not, Be Involved', in this volume, pp. 3—28.
4. President's Commission for the Study of Ethical Problems in Medicine and Biomedical and Behavioral Research: 1982, *Making Health Care Decisions*, U.S. Government Printing Office, Washington, D.C.
5. Rawls, J.: 1971, *A Theory of Justice*, Harvard University Press, Cambridge, Massachusetts.

PART III

FASHIONING LAW AND PUBLIC POLICY

E. HAAVI MORREIM

COMPETENCE: AT THE INTERSECTION OF LAW, MEDICINE, AND PHILOSOPHY

I. INTRODUCTION

In the informed consent setting, our moral interest in competence stems largely from the moral premium we place upon respect for autonomy — roughly, the capacity of a person to shape his own view of the world, and to choose the ways in which he will participate in the world. This capacity, we believe, renders each person morally special, an initiator of thoughts and actions who can bear responsibility for who he is and what he does. Competence, in turn, ordinarily refers to the extent to which a person actually possesses and can exercise such capacities. He who is in full possession of them is considered to be fully competent ([9], [42], [16]).

At the most basic level, we respect autonomy by allowing it to exist — by refraining from killing autonomous persons and by refraining from destroying or damaging that capacity in living persons. On the more everyday level, however, we honor autonomy by honoring its exercise — by deferring to the actual judgments and decisions made by autonomous persons. If the bare possession of the capacity renders a person morally special, only its free exercise can render him responsible, morally an agent. In the medical context, this usually means permitting each patient to develop his own health care goals and to retain considerable control over the ways in which they are pursued.

But if we usually honor autonomy by deferring to decisions, how should we respond to the decisions of those who are not fully autonomous, i.e., whose competence is impaired? Such situations arise often in the health care setting, as a person's ability to understand his situation and to reason through his values to a decision may be impaired by depression, fear, sedation, metabolic imbalance or a host of other factors. A newly paralyzed athlete, for example, may beg to die, unable through his grief to imagine that his life can still be worthwhile.

On the one hand, it is difficult to defend automatic acquiescence to such patients' expressed decisions. After all, those decisions may reflect the competence-impairing factors — the fear or the sedation — more than the person's remaining autonomy. On the other hand, neither can we afford to ignore that decision, for a partial impairment of competence may leave intact some autonomy which still merits respect.

In this paper, we will explore impaired competence both conceptually and normatively. We will consider more fully the relationship between autonomy and decisions, and we will inquire how we should identify and respond to the decisions which issue from impaired competence.

We will begin by considering an initially attractive solution, the appraisal of which will lead us into our conceptual analysis: if the expressed decisions of impaired persons cannot warrant our complete respect, then perhaps we can construct some 'preferred decision' which better reflects the person's autonomy. Deferring to it instead, we could still honor his autonomy even in its absence.

We will see in section II that this approach fails because the autonomous decision has no 'life of its own'. It can only be described procedurally, as *whatever* decision is produced by an autonomous person. Therefore, we can determine whether a decision is sufficiently autonomous to warrant respect only by examining the person himself and his level of autonomy.

This, in turn, requires us to consider what levels and sorts of capacities we wish to require in deciding whether someone is 'sufficiently autonomous'. In section III, we will examine more closely the nature of autonomy and of competence, as applied both to persons and to decisions. We will establish a framework of basic concepts which can then be used to enlighten our practical inquiries: (1) how we should evaluate competence, including how we identify impaired competence, and (2) how we should respond to cases of impaired competence.

Since our primary interest focuses upon the person and his capacities, ideally we should evaluate persons' competence by direct examination. Unfortunately, however, this is usually impossible, for the capacities of autonomy are abilities or dispositions which normally remain unseen until they are exercised in some public performance. Thus, the ability to understand information is indirectly evaluated, e.g., by providing someone with information and requesting him to answer relevant questions. These performances do not always reveal well the abilities which underlie them. A person may choose to conceal his true aims and reasoning, he may be physically or linguistically unable to demonstrate the range of his capabilities, or we as observers may simply misinterpret what we see. The net result is a serious epistemic gap between the evaluation we wish to make and the tools by which we make it. In section IV, we will explore these problems which arise in evaluating competence, and we will see how they prompt us to enlarge the rest of our practical inquiry. We must inquire, not only how to respond to situations of impaired competence, but how to manage our epistemic uncertainties.

In section V, we will investigate the practical management of impaired competence. We will see first that there are, and should be, substantial differences between the standards of competence for which physicians look in clinical practice, and those in the law. Equally substantial differences should characterize the responses to impaired competence in medicine and in law. These differences, I will argue, arise from the differences between the overall goals of medicine and those of law; from the differences between their respective views of human autonomy; between the procedures whereby clinical medicine and the law operate; between the ways in which uncertainties are managed in medicine and in law. We will also discuss the intersection between law and medicine — the ways in which the law may appropriately guide medicine in matters of patient competence, and the ways in which physicians should and should not appeal to law.

II. DIMINISHED COMPETENCE AND THE AUTONOMOUS DECISION

Where a person's competence is (partially) impaired, we are morally uncertain just how to respect whatever autonomy remains. A tempting answer, endorsed or implied in much of the literature, would have us try to identify that particular decision which this individual patient would actually make, were he fully competent, and to observe that 'preferred choice' rather than his expressed choice. In this way, it is thought we can still honor his autonomy by implementing his "true preferences", his "real desires" ([20]), pp. 161, 200, 165; [39], pp. 131, 149).

By what means would we identify this preferred choice? I shall investigate five options and show all of them to be unacceptable. We will thereby understand better what it is for a decision to reflect a person's autonomy, and in turn what this may imply for our moral obligations to respect autonomy.

Option 1: The authentic, well-reasoned, factually well-grounded decision.

The most prominent option stems from a description of the 'fully autonomous decision' which appears in much of the philosophic literature. "A person's decision is autonomous if it derives from the person's own values and beliefs, is based on adequate information and understanding, and is not determined by internal or external constraints that compel the decision" ([9], pp. 44, 114; see also [3], p. 70; [17], pp. 110–111; [20], p. 204; [48], p. 215). A decision is autonomous, that is, only if it is the product of a fully autonomous person who is fully exercising his autonomy. If we wish to identify the autonomous decision which would have been produced by a person whose competence is now impaired, we need only to identify whichever of his values are relevant to his current situation, then apply them to the choices he faces, using the best sort of learning and reasoning process of which he would otherwise be capable. We thereby can morally respect his autonomy even when he is not quite able to exercise it.

There are several major elements in this description of the autonomous decision: (a) the patient's own values (the 'authenticity' requirement), (b) adequate information and understanding, and (c) careful reasoning and deliberation. Let us consider each of these.

(a) 'Authenticity'. This requirement presumes that the patient's stable, long-range values are more autonomous than those which underlie whim or caprice, and that if the patient is unable at the moment to express or to apply his values, we may properly appeal to his most prominent prior values. Indeed, Alan Goldman rests his entire case for justified paternalism on this idea: the "principal criterion is that an individual be acting against *his own* predominant long-term value preferences . . ." ([20], p. 765). If he is, then we may intervene to prevent serious, irreversible harm to that person, but only insofar as this would be designated as a harm by that person's own values. Our paternalism is therefore to protect his true autonomous preferences against his current, bogus, expressed preferences (see also [39], p. 131; [33], p. 473).

Others, while not necessarily sharing this view of justified paternalism, also place great weight on the patient's prior long-range values. Bruce Miller, for instance, suggests that authenticity is one way to define autonomy ([36], p. 24). While they do not believe that authenticity is a necessary condition of autonomy, Beauchamp and McCullough do suggest that we should look for "what the patient would want in a cool and informed moment", his "settled preferences and beliefs, as contrasted with actions and choices motivated by desires and aversions of a momentary, brief, or fleeting duration" ([9], p. 113; see also [41], p. 61n.).

Unfortunately, serious problems arise from an overemphasis on authenticity. First, most people simply do not have a system of values which yields one and only one verdict for each decision. Our value-lives are complex and, when difficult choices are to be made, our values can take us powerfully in several directions at once — with none clearly stronger than the others. Really, this is the essence of finding oneself up against a difficult choice. This incompleteness does not entail a lack of autonomy; it may simply leave room for life's complexity and for personal growth and reflection.

Second, the authenticity requirement seriously misconstrues the nature of human autonomy by ignoring one of its crucial features: the capacity to change one's mind, to re-evaluate and sometimes to reject even long-held values and beliefs. We will discuss autonomy below in more detail. Suffice it for now to say that the bare fact that a person adhered to a value in the past does not entail that he would subscribe to it now, if fully autonomous to reconsider it. Indeed, serious illness can prompt, sometimes even require, major shifts in one's goals and outlook on life ([3], pp. 71—72; [2], pp. 14—15; [6], p. 182; [47], p. 245). To identify and implement that decision which would be most consistent with the patient's past (possibly now invalid) values is therefore not necessarily a good way to honor his autonomy.

(b) 'Adequate information and understanding', and (c) 'Careful reasoning and deliberation'. These two requirements are no more imperative than authenticity. Just as the autonomous person has the capacity to learn and use information for his decisions, so can he autonomously choose to forego information. Just as he can choose to reason carefully, so can he autonomously choose to be arbitrary, even whimsical. Jones can select stock market investments by studying the *Wall Street Journal* and consulting an army of advisors, or he can buy whatever stock's name appears under the left front great toenail of his dog as the dog lies down on the morning paper — and sell whenever the dog wets on the paper (and sell the dog when he wets off the paper). One can autonomously choose to rely on his 'gut instincts', or to follow his spur-of-the-moment impulse rather than to choose always that option which would be produced by a value-programmed computer. To be sure, there are many people for whom careful deliberation, information-gathering, and long-range planning are more of an aberration than a usual decision-making style, such that to require the patient to embrace cautious rationality would be to require him to be inauthentic!

Even more important, many of the values and beliefs upon which even the most highly deliberated decisions rest, are themselves non-rational or even irrational. Our religious beliefs, our most fundamental guiding life-goals, and a wide array of other vital beliefs and values constitute the 'ungrounded grounders' on which we build our view of the world and our own proper place in it ([50], p. 676).

All human conduct is shaped by both conscious and unconscious motivations, reflective and unreflective thought, distorted and undistorted memories, attainable and unattainable aspirations, perceptions which take or do not take reality into account, impulses which brook no delay and those which can first be subjected to experimental thought ([27], p. 112; see also [21], pp. 219–220; [45], pp. 58–59; [33], p. 476).

In sum, there is no good reason to consider a non-reasoned or ill-informed decision, or one which conflicts with prior values, to be intrinsically less autonomous than a well-reasoned, fully informed, value-consistent one.

Option 2: The decision which would be made in the absence of specific internal or external constraining factors.

Instead of the rather abstract approach above, we might construct the 'preferred decision' by identifying whatever specific factors are currently constraining the patient's decision-making and, imagining these to be absent, describing the choice he would then make. Internal factors would include, e.g., fear, depression, pain, metabolic imbalance, etc., and external factors might include family or financial pressures.

Admittedly, the exigencies of illness can have a powerful and adverse impact on a person's thinking. As Cassell has noted, an ill person is not simply a well person with an illness appended like a knapsack ([12], p. 17). Depression, for instance, can render a person so indecisive that he is unable even to be interested in making his own decisions ([44], p. 143; [46], p. 148). Insomnia, anorexia, fear, sedation, and a host of other conditions can seriously dim one's view of his situation and options. Or a lack of life-experience can constrict one's ability to envision realistically the options available to him.

Nevertheless, although these factors can obstruct one's reasoning, they can also sometimes form a legitimate basis for thinking and decision-making. Pain can be a good reason to avoid a particular course of treatment; depression can be a realistic response to a grim situation; "distracting or unsettling emotions" ([17], p. 110) can trigger a necessary re-evaluation of past values; one can autonomously choose to let his family dominate him, or to worry about bankrupting his loved ones.

Such factors, then, do not invariably reduce autonomy. Even granting that they sometimes do, it can be very difficult for an observer to assess the nature and extent of the distortion in the patient's thinking. The observer, after all, suffers a crucial sort of ignorance, namely, the lack of personal appreciation for what it is like to be very ill. From such a distance it may be unwarranted for him to judge that a patient's fears are 'unreasonable' or that his preoccupation with going home is 'inappropriate'. Further, the imaginary removal of assorted thought-affecting factors will not necessarily allow us to construct a decision which is more autonomous than the patient's expressed decision. As serious illness can prompt re-evaluation of long-held values, there is no good reason to view the judgment that the patient would make while not-ill (and not in pain, or afraid, or facing difficult choices) as somehow more 'real' than the judgments he makes while fully experiencing the illness in which and about which the decisions must be made.

Option 3: The decision the patient will embrace in the long run.

If somehow we could identify that choice with which the patient will be happiest when all is said and done, would that decision be more autonomous than the patient's currently expressed choice? This option too is faulty. Our first problem is evidentiary. There is no way of predicting accurately which option the patient will ultimately favor. Autonomy includes the capacity not to change as well as to change one's view, and the autonomous agent is always capable of surprising us.

Second, our definition of 'long run' is likely to be heavily value-laden. We may, for instance, find ourselves judging that the long-run is only reached when the patient finally agrees with us. Also, as we have seen above, there is no convincing reason to suppose that retrospective, non-ill judgments are necessarily more autonomous than those made in the midst of one's illness.

Finally, even if the patient does eventually fulfill our prediction and embrace the outcome we predicted he would, we cannot conclude retrospectively that his original choice was nonautonomous. It could be that the later, rather than the original, choice is the impaired one. Or they may both be autonomous, bridged by an autonomous change of mind. To be sure, even the patient's own subsequent appraisal of his reasoning during the illness can suffer from a serious impairment — his own lack of full appreciation for what it was like to be ill, in pain, afraid. The human memory has a remarkable capacity to forget or repress unhappy experiences. As we noted above, such unhappy factors can sometimes constitute important and legitimate bases for decisions.

Options 4 and 5: The decision that is medically optimal; the decision that an objective, rational observer would make.

These two final options may be discarded quickly. First, there is no assurance in any given case that there will exist some one decision which all reasonable physicians agree is medically best, or one decision which all rational observers agree is objectively best. Reasonable physicians and rational people can disagree about many things, even apparently clear-cut situations.

More importantly, both these options ignore the very thing which we are most anxious to honor: the patient's autonomy, his capacity to formulate and implement his own beliefs and values as he sees fit.

III. A BETTER VIEW: AUTONOMY AND COMPETENCE, PERSONS AND DECISIONS

There is good reason why we have repeatedly failed in our attempt to identify a formula whereby we can describe the particular decision which a given patient would actually embrace if fully autonomous. It concerns the nature of autonomy — the capacity of a person to make his own decisions for his own reasons.

Autonomy is more than the bare ability to select one option over another in accordance with a set of beliefs and values. A computer can do as much. Rather, autonomy renders a human being distinctive by enabling him to re-examine his ideas and preferences and to ask "is this really what I want to do, to believe?" He can examine even his naturally given needs and desires and choose whether to identify with them, shun them, or pursue them. Indeed, the autonomous person can critically reconsider even the very standards by which he makes his judgment — the criteria by which he selects his beliefs and the values that shape his life plans and guide his daily decisions ([33], pp. 470—471; [16], p. 71; [19], p. 156).[1]

Such changes must not be chaotic or pointless, of course. It is equally essential that the autonomous person be able to maintain some stability in his beliefs and values over time. One must form some reasonably orderly and coherent picture of the world and of his place in it. He must be able actually to espouse beliefs and values, and this requires that he recognizes that his beliefs and values can commit him to particular actions and decisions on particular occasions. He must be capable of foreseeing and of either embracing or rejecting the implications of his ideas; of reaching and implementing decisions; of identifying and pursuing goals over time in a fairly orderly manner (see [41], [47], pp. 57—58; [33], p. 470). However — and this is the important point here — this need for overall stability does not dictate anything whatever regarding any particular decision by an autonomous person (any more than the long-range statistical fact that coin flips will yield about 50% heads can tell us how this particular tossed coin will land). Each new situation offers the autonomous person the opportunity to re-consider his prior values and beliefs.

Similarly, it is admittedly essential that the autonomous person be generally rational in his thinking and judging processes. Rationality is crucial to our autonomous ability to identify, analyse and reassess our beliefs and values. But again, this entails nothing with respect to any particular decision. That one must be generally rational does not preclude his autonomously choosing to be spontaneous, impetuous, frivolous or even foolish regarding some particular decision. There is indeed such a thing as a competent, autonomous person choosing irrationally ([13], p. 588).

Because the autonomous *person* is capable by definition of changing his mind at any time about virtually any subject, even of deciding on a given occasion to dispense with caution and rationality, we can only describe the autonomous *decision* procedurally. It is WHATEVER decision is voluntarily, intentionally produced by an autonomous person. It is the product of someone's free exercise of autonomy and, for that reason, cannot be substantively described in advance with any assurance. Because autonomous persons do tend to be stable in their values and beliefs over time, we can of course make predictions as to what they will do, and we may often be right. But it is impossible in principle to construct any reliable formula for describing the autonomous decision independently of the active decision-making of an autonomous person. An autonomous decision can be well- or ill-informed, well- or poorly-reasoned, forward-looking or short-sighted, serious or silly, consistent with one's past views or a radical departure from them. The search for the 'preferred choice' which is intrinsically more autonomous than someone's expressed choice is thus necessarily hopeless.

If the autonomy of persons[2] and autonomy of decisions is as I have described, how then do we describe competence? Elsewhere in this volume, Beauchamp has provided an excellent analysis of competence: its definition as distinguished from its criteria and tests, its relationship with autonomy, and the value issues which enter into the process of setting standards of competence. I largely agree with his discussion, and I shall not repeat his remarks in detail here. I shall briefly summarize them, however, for I do wish to enlarge on his scheme at certain points [8].

Beauchamp has rightly noted that competence and autonomy are different concepts, each with its own grammar. Autonomy features the capacities described above, while competence concerns the ability to do a particular task — any task, from tying one's shoes to designing a space satellite. The requirements of competence in each instance will vary, depending on the task. In the medical context, our use of 'competence' comes very close to our use of 'autonomy', for we are interested in the patient's competence to do one basic task, namely, to make autonomous decisions regarding his medical situation.

Beauchamp has also distinguished between the competence of the person ('competence to') and the objective competence of his actions and decisions ('competence in'). Further, he has suggested that the competently performed task is not necessarily autonomous, nor is an autonomously performed task necessarily objectively competent ([8]), pp. 59—60).

We have in addition seen a distinction between general and specific competence. General competence concerns one's overall ability to function as a rational person, to reason about and to perform the ordinary tasks of daily living. Specific competence, in contrast, concerns the person's ability to perform some particular task, or to function in a particular set of circumstances ([8], pp. 51, 56; see also [28], [42], [21], p. 215). Here we need to enlarge upon previous discussion.

Specific competence must be understood in two very different senses — a basic sense, and a more sophisticated. On the more basic level, specific competence is simply general competence as applied to a particular everyday task. It is the ability to bring the same autonomous reasoning to this task that one can bring to the other ordinary tasks of life. We need to identify this sort of competence because, even if it is appropriate to declare a person to be generally competent, he may nevertheless have serious impairments in a particular limited area and thus lack this specific competence. An otherwise competent person thus may suffer phobias, denial or delusions, for example, which only affect his thinking and decision-making in a narrow range of his functioning. Robertson discusses just such cases in his contribution to this volume ([42], p. 132ff).

In contrast, we also speak of a much more sophisticated sort of specific competence: the ability to perform some fairly complex task, anything from carpentry to surgery, in a reasonably high-quality fashion. Ordinarily this sort of competence requires, not merely the ability to acquire and use information, but the actual possession of relevant information and skills. And one must possess these to such a degree that those who are knowledgeable in that field could agree that he can execute that task satisfactorily, i.e., competently. Thus, one cannot be called competent to fix a malfunctioning truck engine if he only possesses intact the capacity for autonomous functioning in ordinary living, as brought to bear upon this particular area. He must also know plenty about engines and about the tools and techniques with which they can be repaired, and he must have the skills to use this knowledge correctly. Note further, that we must still distinguish between the competence of the person and the competence of his performance. As Beauchamp has noted, one can be capable of performing a task well even if, in a given instance, he does poorly ([8], pp. 56—57; for further discussion of this sort of 'performance competence', see [37], p. 237ff).

Interestingly, when we inquire about specific competence in our ordinary discourse, we nearly always refer to this second, higher-order sense. When we ask whether someone is competent to design a house, for instance, we are not so interested in his general level of rationality as in his level of actual knowledge and skills. Even when we inquire whether someone can make a competent decision, it is usually a question about higher-order specific competence — as where Beauchamp's character, Mr. Zimmer, considers whether he is competent to make his own decisions about the gold and silver markets (p. 57). It is only in a very limited set of cases, most notably the medical context, that we would even consider asking about someone's basic-level specific competence.

Because we are interested in both sorts of specific competence in the medical setting, we must be careful to keep them distinct. Physicians, for instance, sometimes suggest that most patients are not competent to make their own medical decisions since, after all, very few patients have a medical education ([29], [22]). Such a position refers to the higher-order version of specific competence. In contrast, much of the philosophical discussion of informed consent centers only on general competence and on the more basic sort of specific competence.

For our purposes, the important point is this. As we decide what level of competence we wish to require as our standard of 'substantial competence' in any given context, we must decide not only about the degree of competence we seek, but about the type. It may be overly simplistic to suppose, with Beauchamp, that we are choosing threshold points on just one continuum of general competence ([8], p. 56).

Beauchamp suggests that we look for a higher level of general competence in those cases where the risks and consequences of the decisions are more substantial ([8], p. 70), but we need to go further. We may wish, for instance, to require basic-level specific competence in many instances. If Jones has a powerful phobia about surgery which prevents him from even considering surgical options in a rational way, we may wish to regard him as not competent to make a decision about surgery even if he is generally competent in the other areas of his life. In selected situations, we may even wish to require the higher sort of specific competence. Where a patient is invited to participate in an extremely risky research protocol, such as the artificial heart, we may require not only that he be able to understand and reason about his choices, but that he actually possess a considerable amount of information and that he positively demonstrate that he is weighing his options in medically and rationally credible ways.

Note, our aim here is not to offer concrete criteria for determining what sort and level of competence should be sought in each instance, but simply to enrich an already very useful analysis of the concept of competence and of its relationship with autonomy. This being done, we may proceed to matters of both moral and practical import: to discuss the basic factors which should guide our selection of competence standards in medicine and in law and to consider how best to respond to impairments of competence. In order to do this, we must first discuss the actual process of evaluating patients' competence. We will discover that there are serious epistemic problems which will influence substantially our final discussion of competence standards in law and in clinical medicine.

IV. ASSESSING COMPETENCE: CONCEPTUAL AND EPISTEMIC PROBLEMS

We begin by enlarging upon the important distinction between the criteria and the tests of competence. The criteria of competence describe the particular abilities which one must possess relative to the given task, and the minimum degree to which he must possess them. The tests, in contrast, are used to measure the extent to which some individual person actually has those required capacities. Thus, for instance, if we are interested in someone's general competence, our criteria will direct us to look at his ability to understand information (among other abilities), and our tests will attempt to measure the extent to which he actually has this ability.

Competence tests ideally should help us to take a precise, controlled, reliable look at various abilities. The psychiatrists' mental status examination, for example, can help us to assess a range of cognitive and affective functions, assisting us to discern not only the presence of deficits, but their particular nature, extent, probable etiology, and the likelihood and best mechanism of reversibility — whether the problem stems from metabolic imbalance, psychosocial problems, organic brain disease, drug reaction, or from some other source ([25], [21], [26]). A good test must not only provide the information we need, but must also be acceptable in practice: reasonably convenient to use, reliable among various users, etc. ([43]), p. 280).

Our competence tests should also be founded upon conceptually defensible criteria. They should, that is, look for abilities of a type and to a degree that are philosophically defensible with respect to the given task. Thus, when we assess someone's competence to make an ordinary decision such as entering into a simple contract or consenting to routine medical care, we should look only for the abilities of general competence (including perhaps the basic-level specific competence to make the particular decision). We would therefore require, e.g., only that the person be able to understand the relevant information, not that he actually understand it. To require the latter would be to demand a higher order of competence than we could philosophically defend.[3]

Unfortunately, serious epistemic problems arise with respect to virtually any competence testing. Most of the abilities for which we wish to test are not themselves directly observable, but can only be inferred from evidence. We can only view indirectly someone's ability to learn or his capacity to reason, for example, by observing his public performances. These in turn are a function, not only of what he is able to do, but of what he chooses to show us, the way in which we interpret what we see, and a host of other factors which not only are not a part of the abilities we are trying to assess, but which can interfere seriously with that evaluation. Let us consider these problems more precisely.

First, our competence tests rely heavily on person-to-person communication, which is fraught with opportunities for misunderstanding. People do not always attach the same meaning to the same conversation. Their diverse cultural backgrounds, conceptual framework, and practical expectations can color enormously their interpretation of what was said. In one case, for example, a woman dying very painfully of multiple myeloma had refused all medical therapy, including some potentially very helpful palliative care, even food. When asked by her physician "Do you want to die?" she sorrowfully shook her head "No"; when asked by the nurses "Do you want to live?" she sorrowfully shook her head "No". A psychiatrist's consult revealed that she understood the question "Do you want to live?" to mean "Do you want more chemotherapy?"; by the question "Do you want to die?" she understood "Do you want a lethal injection?" The misunderstanding was cleared up, and physician and patient were able to agree on a helpful course of palliative care. Other communication problems abound. A patient may deliberately avoid reporting his thoughts fully and faithfully. He may wish to test his physician's motives; to manipulate the health care team; to elicit sympathy from friends and family, to enjoy the gamesmanship of leading others on; or to secure any of a variety of other hidden agendas. A patient may not have the language or conceptual sophistication to express his beliefs and values explicitly and coherently. He may not feel like discussing certain things with the health care team, or may be embarrassed to reveal his real beliefs and goals. He may not trust the physician, as in the case of an Appalachian man who did not believe the information the physician provided. The physician was black, and the (white, rural) patient could not imagine him to be medically competent. (Further interaction between them, with the help of another physician, eventually led to a trusting relationship.) Communication problems can also be mechanical, as where a patient is intubated and unable to speak, or illiterate, or deaf, or speaks a language foreign to the health care team.

Behavior observation, another testing tool, is typically laden with the same problems. A given series of actions may be consistent with more than one interpretation, and one's judgments may at best be impressionistic.

Further, as we look for 'normal' and 'abnormal' responses, our conceptions of normality, of 'appropriate' and 'inappropriate' behavior, of irrationality as it diverges from eccentricity, can all be highly value-laden, as can our assessments of a person's insight and judgment ([45], p. 54ff).

Beyond this, our tests may not be very sensitive or specific. Our assessments of patients' abilities to understand information, for example, commonly rely on tests of rote factual recall, which may not be the most relevant kind of understanding in matters of informed consent ([35], p. 2474). An over-emphasis on recall "essentially ignores the importance of the patient's immediate response to the information being presented", i.e., the very real and legitimate impact which information has upon our formulation of beliefs and opinions, even where we do not later remember the specific factual bases of those beliefs and opinions ([45], p. 58). Further, our tests also do not always control well the particular way in which the information was presented, what prior beliefs about his disease the patient is bringing to his understanding, the length of time between the original presentation of information and the testing, and other important factors [35].

Finally, we do not understand well the impact which not only the testing process, but the physician-patient relationship itself may have upon patient competence. As Katz notes,

It has been insufficiently recognized that Hippocratic medical practices of keeping patients largely ignorant contribute to making patients appear and act more incompetent than they actually are. The traditional physician-patient interaction makes incompetence a self-fulfilling prophecy ([27], p. 112).

Regarding the phenomenon of patient regression, for instance, Katz wonders about

the contribution which physicians' attitudes toward their patients make toward that regressive pull; nor has much attention been paid to learning whether the regression can be reversed by not keeping patients in the dark, by inviting them to participate in decision making, by addressing and nurturing the intact, mature parts of their functioning ([27], p. 111).

In sum, however defensible our criteria of competence may be relative to the task in which we are interested, the tests with which we measure the relevant abilities are inevitably fallible. We can, of course, take steps to limit that fallibility. When we test for someone's ability to understand information, for example, we can limit the room for error if we first ensure that the person has been provided with plenty of the information we wish him to recall for us, in terms he can understand, from a person to whom he is willing to listen, in an environment with a minimum of stress and distractions, and so forth. If the person refuses to hear information about one subject, e.g., his medical situation, we could undertake a similar sort of testing regarding his ability to comprehend information on unrelated topics ([1], p. 58; [11], p. 251).[4]

Nevertheless, our evaluations will remain flawed. Yet we must undertake such testing. Just as we need to respect the autonomy of those who are capable of making their own decisions, so must we protect and help those who are not. We must improve our testing techniques as best as we can, rely at times on plain common sense and, most important, figure out what, morally, we ought to do with the remaining uncertainties. We need policies in medicine and in law which can help us to decide, not only what courses of action are best regarding those who are quite clearly (in)competent, but also which can offer guidance concerning those cases in which we simply are not sure.

V. IMPAIRED OR DUBIOUS COMPETENCE: MEDICINE AND LAW

The physician-patient relationship is perhaps more likely than virtually any other professional-client relationship to involve diminished capacity on the part of the person seeking help. Fear, pain, metabolic imbalance, or sedation can impede the patient's ability to listen and comprehend; depression or emotional fatigue can interfere with one's usual interest in making one's own decisions; differences in education, culture, or concepts of disease and healing can hamper communication and mutual understanding. The physician's office, and even more so the hospital room, can be an uncomfortable, awkward place, even for the healthy person who has a healthy willingness to speak his mind and ask his questions.

We have already seen that our assessments of patient competence are not always certain and that, where competence is impaired or dubious, there is no 'preferred choice' which we can implement over the patient's expressed choice as a handy route to respecting his erstwhile autonomy. What, then, ought the physician to do? And how ought our public policy, expressed through law and the courts, to address such cases?

We will not find simple solutions with universal applicability. Because each situation is different, we will have to be morally satisfied with rebuttable presumptions — basic values which tell us what to do under clear or ordinary circumstances, how to address our uncertainties, and for what sorts of reasons we might override these initial presumptions. These presumptions thus describe the routine practice of medicine and of law. I shall argue that in important ways, the physician's presumptions should be substantially different from those we embody in our law.

A. *Medicine*

In the routine practice of medicine, the physician must view competence, not as an either/or phenomenon, but as a matter of degree. He must assume that any patient's competence can be compromised at any time in a variety of ways.

I say this, not to encourage physicians to bypass the decision-making prerogatives of their patients, but quite the reverse. The conscientious physician needs actively to seek out and to remedy impediments to the patient's full participation in his health care. He should, where appropriate, 'buy time' for the patient to become accustomed to the facts of his illness and his options; relieve the pain which preoccupies him; help the patient to go home for a few days to alleviate hospital fatigue;

counsel the patient regarding his fears and uncertainties; inquire with the patient's family and friends concerning the patient's emotional needs and the most effective ways of helping him; in short, do anything reasonable and relevant to help this person to face his situation and decisions with as few impediments to clear thinking as possible. The more thoroughly he knows the patient, as through a long term outpatient relationship, the better he will be able to detect slippage in the patient's capacities and the better he will know what measures will restore or enhance those capacities.

As Ackerman and Wear note, this process is hardly *laissez-faire*. The physician must be aggressive at times; whether by challenging the reasoning behind a medically unfortunate choice, or by trying to uncover the hidden agenda which may be leading the patient to make statements which seriously conflict with his own ultimate hopes and aims ([2], p. 16; [3], p. 75; [49], p. 267ff).

Note, the physician is not to engineer a particular decision, or even a particular kind of decision — e.g., a decision that is fully informed and based on careful reasoning. We should reject these goals for the reasons outlined in sections II and III. Rather, the physician's goal should be to place the patient in the position of greatest possible control over his own decision-making process. We require, not that the decision be fully informed, but that full information be easily available. We seek, not that the decision be carefully deliberated, but that the patient be as comfortable and clear-headed as possible, so that he can employ such reasoning if he wishes.

So long as there are time and techniques to enhance the patient's capacity for autonomy, the physician is obliged to entertain a low 'index of suspicion' for impaired competence, and to ameliorate obstacles to autonomy as best he can. When available delay-time is gone and a decision must be made, however, or when it is clear that no other efforts can improve competence further, then the physician's presumptions must shift. He must view competence, no longer on a continuum, but on an either/or basis. Either the patient is substantially competent to make his own decisions, or he is not. And if there is no further time or room for negotiation, the physician should abide by the (substantially competent) patient's expressed wishes, so long as his own professional integrity or his personal moral commitments would not be compromised in doing so. There are several reasons for these presumptions.

First, we have good moral reason to presume that human freedom to conduct one's own life as one chooses should normally be honored rather than constrained. Here is not the proper place for a detailed discussion of the moral value of freedom. We may briefly suggest, however, that freedom concerns more than the value of allowing people autonomously to judge their own interests and develop themselves as individuals ([20], p. 159). And its value goes beyond honoring the basic moral equality and agency of persons by allowing each to shape his own life as he sees fit ([16], pp. 72—73). The value of freedom is found even more fundamentally in the conditions under which we are to maintain a moral order at all in human life, an order in which autonomous people are placed under moral obligations and held accountable for their choices. To be obligated to do X presupposes that one could choose to do not-X; to be accountable for having chosen X presupposes that one could have chosen not-X; freedom of some sort is required in any event. Without further argument on this enormously complex topic, then, I shall assume that we must normally honor persons' freedom, and that he who wishes to do otherwise must meet a strong burden of proof.

Second, the moral primacy of freedom requires that we presume in favor of competence in cases of uncertainty. The burden rests on him who would deny competence. Given the substantial epistemic uncertainties inherent in our evaluations of individuals' competences, this burden may be difficult to meet. If the patient has an explanation to render his apparently-irrational value or belief rational, he may not choose to share it with us. It is he who knows whether he is concealing a hidden agenda, or whether his failure to answer a question is a refusal based on hostility. I do not necessarily wish to argue that people have an 'inner life' which can be known only to themselves, hidden forever from everyone else. On the contrary, I concede readily that sometimes an observer can know a person's thoughts or mental state better than the person himself. I wish only to suggest that not only do we bear the burden to demonstrate that someone is not competent, we also have a burden to show that our evidence itself is not subject to easy overthrow on epistemic grounds.

Finally, there are practical considerations. One does not override a patient's expressed wishes *simpliciter*. One must do so via some practical course of action. The use of physical restraint, deception, forceable sedation, outright coercion, or other such means can often be so intrusive as to constitute a separate moral issue. Thus, normally the physician has little more than his powers of persuasion. Where these do not succeed in changing the patient's views, he has few acceptable alternatives.

Where we have sufficient evidence to judge that a patient is quite clearly not competent, our moral presumption shifts from promoting his autonomy to protecting his interests. If the patient formerly were competent, we should try to identify his interests according to the beliefs and values that he held most strongly prior to his loss of autonomy. We cannot be sure that he would concur with the decision if he were competent, of course. Nevertheless, we really have nowhere else to turn, if we wish to pay our respect to his (former) autonomy.

This presumption is defeasible, of course. Where the patient's medical well-being is at stake, where we have the means to help him significantly, and where we have no good reason to believe he would seriously object to medical intervention, our obligation will surely lie in serving that person's best medical interests for survival and future well-being. Extended discussion here would take us beyond the scope of this paper. Suffice it to say that, if ever there is warrant for unconsented medical interventions, these cases are the surest. Indeed, as Beauchamp and McCullough suggest, arguably these are not genuine instances of paternalism at all ([9], p. 92ff).

Where a physician believes that a patient's competence is seriously impaired and where he needs legal support or guidance in making decisions on the patient's behalf, he may choose to seek judicial action. Here, the standards of competence he invokes must roughly fit the courts, lest much time and effort be wasted.

B. *Law*

If the above presumptions should guide the routine practice of medicine regarding patients' competence, then what standards should our law embrace? In order to answer this question, we must first explore some important differences between law and medicine.

First, the overall goals of medicine differ markedly from those of law, including those particular goals which regard human autonomy. In medicine, our general goals are largely positive: to promote good health, to identify and cure disease, to ameliorate handicaps and return a person to active function, to comfort suffering. Concerning human autonomy, medicine's aim is also positive: to help each person to identify his own health goals, and to help him decide just how he will reach them. More broadly, medicine seeks to enhance each person's overall ability to *be* autonomous by fostering the mental and physical health which are so essential to the larger processes of identifying and pursuing one's life's goals and projects.

The overall aims of law are narrower and typically more negative: to protect people against harm by forbidding citizens to treat each other in certain ways; to secure individual freedom by proscribing certain sorts of personal interactions; to secure general social order by outlawing specified disruptive activities; to resolve disputes when they cannot be peaceably resolved elsewhere. Routinely, our legal system presumes to guide citizens' action with a minimum of its own intervention, becoming directly involved only when something goes wrong, when its rules of guidance have somehow failed or been violated. Regarding human autonomy, the law's aim is again more negative than positive: it seeks, not to guide citizens toward self-fulfillment, but to preclude certain sorts of interference with the individual. Rather than the supportive model of autonomy-enhancement which we recommended for medical practice, the law's view of autonomy is the more adversarial 'don't tread on me' model.

Second, medicine and law differ procedurally. Law must focus more on what is common among cases than on what individuates them. In order to generate a set of rules which is fair, consistent, and reasonably predictable, the law must subsume a wide variety of cases under a limited number of general rules. In adjudicating these rules one must decide, in each instance, whether it is a case falling under law X, under law Y, etc., or under no law at all. Only a limited number of individuating factors can be heeded, for each case must be subsumed under some generality. If each instance were treated as unique, utterly different from all other cases, we would not have law at all, but a series of *ad hoc* unprincipled edicts. As a result, law is necessarily somewhat vague — a blunt instrument incapable of addressing all of the peculiarities of an individual case.

Medicine, in contrast, is primarily concerned with the individual case. As a scientific enterprise, medicine does of course seek useful generalizations about the normal and pathologic functions of the human organism. Its chief clinical focus, however, is ultimately upon the individual patient, who may defy all our attempts to subsume him under categories and commonalities. Thus, a generalization is used only to the extent that it fits the case at hand. The fact that ninety-eight per cent of all patients with disease D will benefit from treatment T is of little interest if there is good reason to suppose that patient P's case of D is among the two per cent which will not benefit from T. Whereas in law the individualities of the case are attended to or ignored in order to secure a reasonable fit with the relevant rule, in medicine it is the generalities which are heeded or ignored according to their relevance to the individual case.

These differences of goal and of procedure between law and medicine have important implications for our view about the proper presumptions which the law should embrace regarding the competence of patients to decide about their own medical care. In particular, they support two presumptions: (1) our legal standards should assume patients to be competent in the absence of clear, convincing evidence to the contrary; and (2) competent patients should be allowed to negotiate with their physicians the ends and means of their health care with a minimum of interference. In sum, the law should neither permit nor itself engage in undue interference with individuals' health care decisions. Let us explore these two presumptions in more detail.

(1) The presumption of competence.

There is already in our law the strong presumption for competence.

The well-settled rule concerning the presumption of sanity is that the law will presume ... competency rather than incompetency ..., that every man is capable of managing his own affairs and is responsible for his own acts, and all proceedings testing the competency of a person to perform a certain act will start with a presumption of competency, which can be relied on until the contrary is shown ([32], p. 1440).

This presumption has two elements: a substantive element wherein we require only a rather modest level of general competence in order to judge someone capable of handling his own affairs,[5] and a procedural element wherein we place a strong evidentiary burden upon anyone who would claim a person to be incompetent.

The substantive standard requiring only minimal competence is supported primarily by the now-familiar moral value of freedom. The legal system, as chief guardian of citizens' freedom, is also the chief vehicle whereby freedom can be decisively taken away. A declaration that someone is even partially or temporarily incompetent means that others may assume a large measure of control over his affairs. Therefore, a legal system interested in preserving human freedom should be systematically suspicious whenever one person seeks to declare another incompetent and to assume power over him, even for allegedly benevolent reasons. We must be concerned not only with such high-minded encroachments on liberty, but also with the serious abuses to which declarations of incompetence may be subject (witness the political uses to which judgments of mental illness have been put in the Soviet Union). The more rigorous our standards for certifying incompetence, the less license there will be for intervention. Freedom is, of course, not the only value here at stake. We do have obligations to guard the welfare of those who are unable to look after themselves. Nevertheless, in the end the burden must fall squarely on him who wishes to judge another incompetent, rather than on the 'accused' to prove himself competent.

The procedural element stems from the fallibility of our techniques for evaluating competence as discussed in Section IV. In the law, it is important that not only our rules, but our mechanisms for adjudicating them, be open to public inspection and discussion. That is, just as the laws themselves must be public if people are to be able to know and follow them, so must the criteria and evidence whereby we judge application of the laws be public, so that we can consider whether the laws are being applied properly. Speculation, hearsay, and hunches cannot be sufficient to judge someone incompetent. And yet, in many of the cases involving situations of borderline or partially impaired competence, our evidence is of just this character — more suggestive than decisive. A family member may say, "he's just not himself today", or a physician may wonder just how much the patient's mild hypercalcemia is affecting his thinking. Such situations are morally worrisome because we do have some informal reasons to believe that the person is not fully able to engage in autonomous decision-making, and we are uneasy about taking his expressed choices at face value. Our point here is that these cases generally do not belong in the legal system. There is virtually no way to demonstrate satisfactorily the existence and extent of such impairments and, even if we could, our threshold for declaring incompetence must not be set at such demanding level. Such borderline cases ought best to be resolved outside the law.

(2) The presumption of non-interference.

We have already amply discussed reasons supporting the conclusion that competent patients and physicians should be free to negotiate health care decisions with a minimum of legal interference. Public welfare may sometimes override individuals' liberty, as where contagious diseases are subject to reporting and possible quarantine. But in the absence of powerful reasons to the contrary, citizens should be allowed to define and pursue their own good.

C. *The Intersection of Law and Medicine*

This legal-philosophic discussion has important implications for medicine. It is physicians, after all, who are especially likely to recognize impairments of competence, and it is they who must regularly decide which cases of diminished competence they wish to bring to legal adjudication. Others may raise competence questions, of course, but physicians confront the issue considerably more often than others. The chief lesson physicians may take from our legal discussion is this: medicine should appeal to law only sparingly in the management of patients whose competence is impaired or dubious, even though, under certain circumstances, such an appeal is warranted.

There are both philosophical and practical reasons for minimizing the resort to law. We have already seen that the law should be and is generally reluctant to declare persons incompetent or to intervene in their medical decision-making. One form in which this reluctance appears is the vagueness with which statutory definitions of competence are typically written, as they use such vacuous phrases as "of sound mind", and "(un)able to manage one's affairs" ([25], p. 365).

This definitional vagueness, indeed the vagueness we find throughout the law, can be troublesome at times, for we may find ourselves uncertain, wondering which laws apply to our proposed conduct and in what manner. Nevertheless, vagueness brings flexibility. The less precisely the law tells us what to do and when, the more we are able to use our own judgments. As soon as we request a judgment from the courts, the judiciary must provide a definitive yes-or-no answer, enforced with its weighty authority. Our maneuvering room has been closed. Once the patient is declared legally competent, we ordinarily must defer to his wishes at face value — not necessarily the best way of honoring his somewhat-impaired autonomy, as we have seen. And if he is declared to be legally incompetent, then he may well lose control even over those areas in which he could exercise satisfactory decision-making. This is an offense to his remaining autonomy.

What this means for medicine is that, so long as physicians operate legitimately within the limits of good legal and moral practice, they would normally do best to take advantage of the law's maneuvering room to individualize their approach to each patient with impaired competence. A good physician-patient relationship is not a contest of your-will-or-mine, which can easily develop if the adversarial proceedings of the law are invoked. Rather, physician and patient should work together to explore in the greatest possible depth the patient's view of his situation and his aims in resolving it. The physician is ordinarily much better suited than the judge to ease the patient's fears, to adjust his metabolic balance, to relieve the pain which preoccupies him, to negotiate a compromise with the patient who refuses the medically optimal course, to uncover and correct the misconceptions which may be fuelling a depression. And the patient is ordinarily better suited than his attorney to show the physician why he feels as he does, to offer the explanation which can render a puzzling preference intelligible. Such communications require trust and a mutual willingness to work together toward a common goal — not the routine intervention of third parties.

Practical considerations also support the minimization of medicine's appeal to law in resolving competence cases. Because our legal definitions and statutes are vague, courts typically solicit expert testimony as they decide what, in a given instance, shall be understood by the term 'competent' and whether a given individual is competent. Those experts are, indeed, physicians — sometimes the very physicians who have asked the courts to settle the question in the first place. Thus, as the physician asks the courts whether a patient is competent, the courts may simply return the question to the physician ([25], p. 364). Further, the more burdened the judicial system is, the less likely it is that the patient will receive a careful and independent analysis — and the more likely that the decision will be "so perfunctory and/or deferential to the professional expertise of the providers that the role of the courts amounts to little more than pro forma ratification what was readily apparent to health professionals" ([41], p. 175). Further, the appeal to judicial intervention can be costly and time-consuming for all concerned. And an official declaration of incompetence can cause social and psychological injury to some patients, even where that declaration is technically accurate ([21], p. 222).

Does this mean that legal standards should be largely ignored, and judicial procedures spurned? No. While I do argue that physicians should be very cautious in their resort to legal machinery, that machinery nevertheless serves important functions, and should be readily invoked when the occasion warrants.

First, our legal standards set the basic norms which are to guide the physician-patient relationship. They emphasize that each person must be presumed to be in charge of his own life and future, and that others ought not to interfere unduly, even where his decisions seem quite foolish. Such guidance is entirely appropriate. One of the most important functions of law in a free society is to set some limits on the ways in which people may intrude upon one another, even where such intrusions are benevolently motivated.

Further, some of this basic legal guidance can only be obtained by actually invoking the judicial process. Although our definitions of competence and some of these guiding norms are written into statute,[6] ours is nevertheless largely a common law system, in which the real substance of the law emerges only through case and precedent. Thus, an important portion of the law's legitimate guidance function can only be served when citizens are willing actually to submit their questions and disputes before the courts.

A second crucial function of law is to protect citizens — to enforce its guidance both directly by coercing compliance with its verdicts and also indirectly, through the deterrent force of its enforcement mechanisms and through the authority and power for which its citizens have respect. Normally in the medical context, this protection is not to avert harms which arise from evil intentions. It is more commonly to shield competent but vulnerable patients from overzealous professional benevolence. And it is to ensure that others, such as family members, also do not inordinately dominate that patient's personal decisions.

Note also, however, that such protection is also offered to health care professionals and others. Just as the law indicates the sorts of situations in which intervention is not acceptable, so does it specify conditions under which it is, as where an incompetent patient requires emergency treatment. If the physician acts within the appropriate guidelines, he may be protected from various subsequent legal reprisals.

Legal protection is also available to remedy past injustices and to reverse ongoing injustices. Anyone who perceives that he has been offended — patient, professional, family member — may raise his concerns before the courts. An elderly person, heretofore treated as incompetent, for example, might win a declaration of competence that could restore him to rightful control over his own affairs.

Third and finally, there are times when an appeal for judicial intervention is simply useful. There are times when we need an official declaration that someone is incompetent, for instance, as where such a declaration is prerequisite to making suitable financial arrangements for someone's care. In other cases, judicial intervention can produce clarity out of chaos, as where a judge appoints one member of a large, squabbling family to make all decisions for the incompetent patient.

Thus, the law can lend guidance, protection, and useful services. Admittedly, there are respects in which our laws as they currently stand could be improved. Our laws have tended, for example, to regard competence too globally, to assume that incompetence to do X entails incompetence to do Y [8]. And courts are not always consistent or predictable in the standards they require for competence ([43], [5]). Or they may regard competence too narrowly, as consisting only in the cognitive 'ability to understand', without reference to such equally important abilities as the capacity to produce at least marginally rational reasoning processes. Further, as Robertson notes, it is wrong to assume that certain consequences follow automatically from a declaration of competence and certain other consequences, from a declaration of incompetence [42]. Also, our judiciary sometimes works from too narrow a set of options, e.g., turning over all decision-making power to a guardian in cases where the patient is only partly impaired.

Some of these weaknesses can be at least partially remedied. We can sometimes establish limited guardianships (giving the guardian decision-making power only in specified areas) instead of the usual all-or-none guardianship [21]. We can enrich our legal definitions of competence to acknowledge that the concept is more complex than simple 'ability to understand'. And we can recognize that competence judgments need not be so global as we sometimes make them.

Nevertheless, however much we improve it, our law will necessarily remain a rather blunt instrument. And so although it plays a crucial role, for the reasons outlined throughout this section we must be judicious about when and where we invoke legal proceedings.

VI. CONCLUSION

In much of the philosophic literature on informed consent, we are told "if the patient is competent, here are the values which should prevail" and "if he is incompetent, here is what to do . . ." There has been little appreciation of the enormous variety of ways and degrees in which competence may be compromised, nor recognition of the important implications this diversity may have.

Situations in which a person's competence is only partially impaired are perhaps the most difficult of all because the person retains some, but not all, of his capacity to make autonomous decisions. Given that our moral aim is generally to honor autonomy, and given that normally we do this by deferring to persons' expressed decisions, we are puzzled in these cases by the relationship between the patient's autonomy (whatever remains of it) and the wishes he expresses. Those wishes may not genuinely reflect his autonomy, and so a blind deference to such preferences taken at face value is not morally acceptable, particularly where doing so would produce seriously adverse health consequences. Neither can we safely ignore the patient or his wishes, for *ex hypothesi* he does retain at least some ability to think and decide for himself.

In this paper I have argued that the responses of medicine and of law to these situations should be markedly different. Our law should provide general guidance, protection and, in select instances, definitive resolution and enforcement of disputed questions. It tells us that, as a rule, we must presume persons to be capable of handling their own affairs and that we must honor that capacity. On the everyday level, however, we must remember that law is an adversarial process which is by nature incapable of responding fully to the important peculiarities of each case. In those cases of partial or borderline impairment of competence in which we are especially interested, routine resort to the courts could work far more mischief than benefit.

Much better suited to these situations is a warm, mutually respectful physician-patient relationship. Whereas the law should ordinarily presume patients to be competent, the physician in routine clinical practice must be constantly alert for impairments of competence — not for the purpose of usurping the patient's decision-making authority, but in order to enhance the patient's ability to participate as fully as he wishes in his own health care. In other words, there are many instances in which a physician will rightly regard as not-so-competent a patient whom the law, equally rightly, would regard as fully competent. Whereas the law's aim is to protect citizens from improper intrusions, the physician's aim is to help his patient to identify important goals and the means that shall be taken toward them, and to help him to learn to live with those health problems over which no one has control.

Ultimately law and medicine alike aim both to help and to respect people. But with their rather different outlook and their very different procedures, each must serve this aim in its own way — and neither must presume to tread too far into the other's domain.

University of Tennessee
College of Medicine
Memphis, Tennessee

NOTES

[1] Of course there are other abilities required for autonomy, including the abilities to understand and manipulate information, to deliberate rationally, to draw conclusions and initiate action when warranted, and so forth. I shall not discuss these here, for that has been done elsewhere ([9], [37], [14]). Our focus is on those more subtle factors which produce partial or borderline impairments.

[2] We may note that this description of the autonomous person differs from that of S. I. Benn. According to Benn, a person is only autonomous to the extent that he actually exercises his capacities for formulating and reassessing his values and beliefs. He who only possesses but does not employ such capacities is best called autarchic ([10], p. 124); see also Miller's fourth definition of autonomy, "moral reflection", ([36], p. 25). If my account is correct, however, we must disagree with Benn. Autonomy is the capacity not only to choose one's basic standards of value and belief, but the ability as well to choose just when one shall employ such standards, including when one will ignore them. This is a part of the ability to reconsider and sometimes to change one's standards.

[3] Interestingly, in the literature on informed consent, there is a striking and pervasive failure to distinguish properly between the criteria and the tests of competence. Even though Beauchamp expressly draws this distinction (pp. 64—65), his actual discussion of competence tests repeatedly confuses the two, as he refers to "inabilities" as "tests" (p. 65). As should be evident by now, the (in)abilities in which we are interested are capacities for which we test, and are not themselves mechanisms of measurement. The same flaw appears in the work of the Pittsburgh Group as they speak, e.g., of "ability to understand" as if it were a test of competence rather than the very capacity for which we test ([43], p. 280ff.; see also [45], p. 53ff.; [34], p. 29; [4], p. 952ff.). It is important to us to guard against this error because, to the extent that we speak interchangeably of abilities and the tests with which we measure them, we may wrongly assume that our tests reveal those abilities much better than they actually do. We may, that is, lose track of the major epistemic problems which taint our assessments of persons' competence capacities. These problems are discussed in the text, below.

[4] We can also make use of what I shall call 'suspicion-triggers'. Unlike *bona fide* tests of competence, which (should) provide fairly reliable evaluations toward a conclusion that a person is (in)competent, these triggers do not themselves provide concrete evidence toward a conclusion that someone is (in)competent. At most, they indicate that something may be amiss, that we need to look more closely at the situation. The following are five important examples of such triggers.

(1) *The presence of an 'obviously false' belief.* A major discrepancy between the facts of a situation and the patient's description of it — or between his interpretation of his situation and the normal interpretation of it — requires explanation. That explanation need not be reasonable, so long as it is reasonably intelligible ([33], p. 479). A patient who denies that her black, foul-smelling foot constitutes a health problem, for instance, may have religious beliefs which define 'health' not in physical terms, but in spiritual. With such an accounting, her 'false belief' does not necessarily indicate incompetence.

(2) *A flagrantly unreasonable decision.* Similarly, a decision that is radically at odds with one's best medical interests requires an intelligible account. With such an account, however, our suspicions of incompetence are eased. Now we can see that Roth *et al.*'s second 'test' for competence, 'reasonable outcome', is not a test at all, but a suspicion-trigger. The bare fact that a patient came to a particular (un)reasonable conclusion does not entail anything about his competence ([43], p. 280).

(3) *Major pre-existing cognitive, rational or affective deficits.* Patients with serious mental illness, retardation, organic brain disease, significant depression, etc., may or may not be incompetent regarding a particular decision, but they do at least prompt close scrutiny ([21], p. 217).

(4) *Major divergence with stable past beliefs or values: inauthenticity.* Such changes of belief or value can reflect an autonomous change-of-mind, or a serious competence problem. Only further investigation can settle the question. The patient must be able to recognize that he is functioning according to different beliefs or values, and to provide a suitable accounting ([14], p. 54; [36], p. 24; [37], pp. 235, 241—242).

(5) *Extreme vacillation or high likelihood of imminent change of mind.* A high degree of ambivalence could represent a genuine conflict between powerful, incompatible values, or a flawed ability to sort out one's thoughts and reach a conclusion.

Suspicion-triggers such as these five are useful, because they prompt us to examine the patient's competence more closely, and because they remind us at the same time of the fallibility of our own examinations. Ultimately, we cannot avoid epistemic gaps as we evaluate the competence of individuals, for we are inferring from the seen to the unseen. That someone does not display an ability does not entail that he cannot.

[5] We argued above that in certain areas the law might rightly require a more rigorous standard of competence than it normally does. That claim does not conflict with the point we have just made. Together they mean that our normally modest standards of competence may on occasion be raised, but only with strong justification.

[6] Such statutes, of course, will differ from state to state.

BIBLIOGRAPHY

1. Abernethy, V.: 1984, 'Compassion, Control, and Decisions About Competency', *American Journal of Psychiatry* **141**, 53—58.
2. Ackerman, T.: 1982, 'Why Doctors Should Intervene', *Hastings Center Report* **12**, 14—17.
3. Ackerman, T.: 1984, 'Medical Ethics and the Two Dogmas of Liberalism', *Theoretical Medicine* **5**, 69—81.

4. Appelbaum, P. S. and Roth, L. H.: 1982, 'Competency to Consent to Research', *Archives of General Psychiatry* **39**, 951–958.
5. Appelbaum, P. S. and Roth, L. H.: 1982, 'Treatment Refusals in Medical Hospitals', in *Making Health Care Decisions (Volume Two: Appendices)*, U.S. Government Printing Office, Washington, D.C., pp. 411–477.
6. Baumgarten, E.: 1980, 'The Concept of Competence in Medical Ethics', *Journal of Medical Ethics*, **6**, 180–184.
7. Beasley, A. D. and Graber, G. C.: 1984, 'The Range of Autonomy', *Theoretical Medicine* **5**, 31–41.
8. Beauchamp, T.: 1991, 'Competence', in this volume, pp. 49–77.
9. Beauchamp, T. and McCullough, L. B.: 1984, *Medical Ethics*, Prentice-Hall Inc., Englewood Cliffs, New Jersey.
10. Benn, S. I.: 1976, 'Freedom, Autonomy and the Concept of Person', *Proceedings of the Aristotelian Society* **76**, 109–130.
11. Burra, P., Kimberley, R. and Miura, C.: 1980, 'Mental Competence and Consent to Treatment', *Canadian Journal of Psychiatry* **25**, 251–253.
12. Cassell, E.: 1977, 'The Function of Medicine', *Hastings Center Report* **7**, 16–19.
13. Culver, C. M., Ferrell, R. B. and Green, R. M.: 1980, 'ECT and Special Problems of Informed Consent', *American Journal of Psychiatry* **137**, 586–591.
14. Culver, C. M. and Gert, B.: 1982, *Philosophy in Medicine*, Oxford University Press, New York, pp. 20–41, 42–63.
15. Dworkin, G.: 1976, 'Autonomy and Behavior Control', *Hastings Center Report* **6**, 23–28.
16. Dworkin, G.: 1982, 'Autonomy and Informed Consent', in *Making Health Care Decisions (Volume Three: Appendices)*, U.S. Government Printing Office, Washington, D.C., pp. 63–81.
17. Feinberg, J.: 1971, 'Legal Paternalism', *Canadian Journal of Philosophy* **1**, 105–124.
18. Frankfurt, H. G.: 1971, 'Freedom of the Will and the Concept of a Person', *The Journal of Philosophy* **68**, 5–20.
19. Gaylin, W.: 1982, 'The "Competence" of Children: No Longer All or None', *Journal of the American Academy of Child Psychiatry* **21**, 153–162.
20. Goldman, A.: 1980, *The Moral Foundations of Professional Ethics*, Rowman and Littlefield, Totowa, pp. 156–229.
21. Gutheil, T. C. and Appelbaum, P. S.: 1982, *Clinical Handbook of Psychiatry and the Law*, McGraw-Hill Book Co., New York.
22. Inglefinger, F. J.: 1982, 'Informed (But Uneducated) Consent', *New England Journal of Medicine* **287**, 465–466.
23. Jackson, D. L. and Younger, S.: 1979, 'Patient Autonomy and Death With Dignity', *New England Journal of Medicine* **301**, 404–408.
24. Jackson, D. and Younger, S.: 1982, 'Autonomy and the Need to Preserve Life', *Hastings Center Report* **12**, 44.
25. Janicak, P. G. and Bonavich, P. R.: 1980, 'The Borderland of Autonomy: Medical-legal Criteria for Capacity to Consent', *The Journal of Psychiatry and Law* **8**, 361–387.

26. Kaplan, H. I. and Sadock, B. J.: 1981, *Modern Synopsis of Comprehensive Textbook of Psychiatry/III*, 3rd edition, Williams and Wilkins, Baltimore, Maryland.
27. Katz, J.: 1981, 'Disclosure and Consent in Psychiatric Practice', in C. K. Hofling (ed.), *Law and Ethics in the Practice of Psychiatry*, Brunner/Mazel, New York, pp. 91–117.
28. Knight, J. A.: 1991, 'Judging Competence: When the Psychiatrist Need, or Need Not, Be Involved', in this volume, pp. 3–28.
29. Laforet, E. G.: 1976, 'The Fiction of Informed Consent', *Journal of the American Medical Association* **235**, 1579–1585.
30. Leikin, S.: 1982, ' "Minors" Assent or Dissent to Medical Treatment', in *Making Health Care Decisions (Volume Three: Appendices)*, U.S. Government Printing Office, Washington, D.C., pp. 175–191.
31. Lo, B. and Jonsen, A.: 1980, 'Clinical Decisions to Limit Treatment', *Annals of Internal Medicine* **93**, 764–768.
32. Lockwood, R. H.: 1969, 'Mental Competency of Patients to Consent to Surgical Operation or Medical Treatment', *American Law Reports Annotated* 3d. Series **25**, 1439–1443.
33. Luban, D.: 1981, 'Paternalism and the Legal Profession', *Wisconsin Law Review*, 454–493.
34. Meisel, A.: 1981, 'The "Exceptions" to Informed Consent', *Connecticut Medicine* **45**, 27–32.
35. Meisel, A. and Roth, L. H.: 1981, 'What We Do and Do Not Know About Informed Consent', *Journal of the American Medical Association* **246**, 2473–2477.
36. Miller, B. L.: 1981, 'Autonomy and the Refusal of Lifesaving Treatment', *Hastings Center Report* **11**, 22–28.
37. Morreim, H.: 1983, 'Three Concepts of Patient Competence', *Theoretical Medicine* **4**, 231–251.
38. Murphy, J.: 1979, 'Therapy and the Problem of Autonomous Consent', *International Journal of Law and Psychiatry* **2**, 415–430.
39. Muyskens, J.: 1982, *Moral Problems in Nursing*, Rowman and Littlefield, Totowa.
40. Ozar, D.: 1984, ' "Patients" Autonomy: Three Models of the Professional – Lay Relationship in Medicine', *Theoretical Medicine* **5**, 61–68.
41. President's Commission for the Study of Ethical Problems in Medicine and Biomedical and Behavioral Research: 1982, *Making Health Care Decisions*, U.S. Government Printing Office, Washington, D.C.
42. Robertson, J.: 1991, 'The Geography of Competence', in this volume, pp. 127–148.
43. Roth, L. H., Meisel, A. and Lidz, C. W.: 1977, 'Tests of Competency to Consent to Treatment', *American Journal of Psychiatry* **134**, 279–284.
44. Sherlock, R.: 1983, 'Consent, Competency and ECT: Some Critical Suggestions', *Journal of Medical Ethics* **9**, 141–143.
45. Tancredi, L.: 1982, 'Competency for Informed Consent', *International Journal of Law and Psychiatry* **5**, 51–63.
46. Taylor, P. J.: 1983, 'Consent, Competency and ECT: a Psychiatrist's View', *Journal of Medical Ethics* **9**, 146–151.

47. Thomasma, D.: 1983, 'Beyond Medical Paternalism and Patient Autonomy', *Annals of Internal Medicine* **98**, 243–248.
48. Watson, G.: 1975, 'Free Agency', *The Journal of Philosophy* **8**, 205–220.
49. Wear, S.: 1983, 'Patient Autonomy, Paternalism, and the Conscientious Physician', *Theoretical Medicine* **4**, 253–274.
50. Werner, R.: 1983, 'Ethical Realism', *Ethics* **93**, 653–679.

JOHN A. ROBERTSON

THE GEOGRAPHY OF COMPETENCY

On maps of the Western states, the zig-zagging but nonetheless clear line of the continental divide indicates to which oceans the streams and rivers arising there will run. The concept of competency is a continental divide of sorts in the ethics and public policy of medicine. Patients on one side of the watershed of competency get to make treatment decisions, while those on the other side do not.

Brief experience with competency determinations, however, reveals that law, ethics, and medicine are less neat and more complicated than the sure ridgeline of the continental divide. Competency also appears to be a binary concept, but on closer examination its binariness dissolves and vanishes, or remains tufted and turretted, hardly the clear guiding line for action expected. This paper takes a brief hike through the terrain of competency in order to provide a more accurate map of its geography.

I. THE VALUE OF AUTONOMY

The apparent watershed nature of competency has emerged from a consensus in law, ethics and public policy about value choices in bioethical contexts. Most issues of bioethics concern issues of autonomy, beneficence, or justice [43]. Autonomy has the most wide-spread support and the clearest definition. (Indeed, notorious violations of autonomy in human subject research and in failures to honor religious objections to medical treatment originated the field of contemporary medical ethics in the 1960s.) There is now a widespread consensus in medical ethics and law that the autonomy of the individual is a first-order principle, a central value that can be overridden only by very strong reasons.

The common law system has always recognized this value in the law of assault and battery.[1] In the 1960s and 1970s, the legal protection for autonomy in medical decisions was extended in several important ways. Informed consent doctrines in malpractice law enhanced protection of autonomy by assuring the information needed for a knowing autonomous choice.[2] *Roe v. Wade* and its progeny established reproductive autonomy.[3] Finally, state courts explicitly recognized that personal autonomy included the right to refuse medical treatments that could extend a person's life. Although derived from civil law doctrines of battery and informed consent, constitutional protection for this right was found in the liberty clause of the fourteenth amendment, making it a fundamental right which the state could override only by showing a compelling need for treatment ([12], [3], [20]).

The legal protection granted to autonomy grew out of a substantive moral position about the worth and meaning of respect for persons. Basically, that position is that respect for persons — treating them as ends rather than as means merely — entails that we treat them as autonomous persons who make choices, as persons with the power and right of self-determination across most affairs of life. This value position rests on the notion that persons generally do have the capacity and ability to make choices, and generally exercise it. By its very terms, however, the value of autonomy does not extend self-determination to non-choosing persons, or to persons who lack the capacity for choice. Competent persons get to decide for themselves. Children, the retarded, the mentally ill, and similar persons do not. Persons falling on the incompetency side of the line still deserve respect as persons, but respect for them does not mean that we honor their preferences or other expressions of choice.[4]

II. THE DOMAIN OF COMPETENCY

Discussions of competency in this volume as in most settings usually focus on decision-making with regard to adults who refuse treatment. The person appears to be competent, but refuses a medical therapy that the doctor thinks is desirable. The discussion then concerns whether the person is competent, how competency is determined, and the consequences of a judgment of incompetency.

Respecting refusals of treatment is, of course, important, involving a large portion of the situations in which competency is relevant. But we should not lose sight of the many other situations in which questions of competency arise, for they overlap and intertwine in ways that influence the issues in any area. A brief discussion of three other contexts in which competency arises will show the mutual influence.

First, competency issues involving adults extend to many other issues than refusing conventional acute treatment. They can and do arise over an adult's competency to accept treatment. More obviously in this regard, they arise with the issue of adult consent to an experimental therapy. Federal regulation of human subject research through the Institutional Review Board (IRB) system is based on the moral principle of autonomy. With a few exceptions, only persons who knowingly and freely consent to research may participate as subjects. The competency or incompetency of a person will then determine what set of research rules apply, and in most cases, whether the research will be done at all.[5] Thus Barney Clark's consent to experimental implantation of the artificial heart was not valid unless he was competent to consent.[6] Attention to the meaning, scope, consequences, and procedures for dealing with competency in the experimentation context will be relevant in other contexts.

Questions of competency also arise with adults in non-acute situations that cast light on competency in refusals of acute care. One such situation arises when parents of a mentally retarded woman request a physician to sterilize her. While there have been abuses involving sterilization of the retarded, an intermediate position has developed in some states allowing sterilization of the incompetent retarded if the sterilization can be shown to be in her interests (see, e.g., *In re Grady* (1981) [8] and *In re Mary Moe* (1982) [10]). In states without that rule, such as Wisconsin, a woman could not be sterilized unless in fact competent (see, e.g., *In re Guardianship of Eberhardy* (1981) [9]). Again, the meaning of competency in this setting may influence its meaning in other settings.

Involuntarily committed mentally ill patients also face issues of competency when they refuse psychotropic medication thought necessary by their doctors. It is now well recognized that involuntarily committed mental patients, though mentally ill, may remain competent to do many things, including making medical and psychiatric care decisions. Recent cases such as *Rennie v. Klein* (1978) [18] and *Rogers v. Okin* (1980) [19] have recognized that the loss of liberty entailed in involuntary hospitalization does not necessarily deprive the person of all other liberties, and thus have established the right of such patients when competent to refuse medications for their mental illness [35]. While debate continues over the circumstances which justify overriding a competent refusal, the competency line here plays a similar watershed role. Again the tests, meaning, and procedures for determining competency in this setting will influence and illuminate competency issues in the adult refusal of acute care cases.

Finally, questions of competency arise with older children across the range of issues raised with adults, including refusal of life-saving acute care, sterilization, experimentation, and treatment of mental illness. Generally we talk about the rights of competent adults and tend to overlook the extent to which 'children' are also competent. On the 'understand-the-nature-and-consequences' test, most teenagers, and many younger children, would be competent to make medical decisions and thus arguably should have the same rights as adults. Indeed, the law has increasingly recognized the mature minor's right of self-determination in health care, and the U.S. Supreme Court has recognized their right of autonomy in abortion and birth control decisions [50]. If they have a right to have or refuse an abortion, arguably they should have the same right of privacy as competent adults to refuse medical care.

The issue of the competency of minors to refuse medical treatment is of great importance with critically ill children who find the rigors of chronic critical illness burdensome. If an adult may discontinue dialysis and die, may not a 16-year-old who is equally burdened ([51], pp. 93—94)? If so, should a 12-year-old decide whether to take the 50% risk of dying from graft vs. host disease arising in a bone marrow transplant, as David, the Houston 'bubble-baby' did and lost ([34], [37])? Should we respect the wishes of 12-year-old Pam Hamilton, who, along with her parents, refuse established chemotherapy on religious grounds [38]? Although parents may not make martyrs out of their children by denying them life-saving treatment, should not Pam, if she is competent, be free to make a martyr out of herself? The concept of competency, tests for determining it, and rules about its consequences with adults will influence the decision here, which in turn will influence it in adult contexts as well.

Our discussion of competency needs to be broadened to include these other contexts. The impact, procedures, meaning, and consequences of competency across a range of areas is essential to understand it in any particular area. With this broadened perspective in mind, let us now turn to the problematic aspects of the competency judgment.

III. THE DIFFICULTY OF DEFINING AND APPLYING THE CONCEPT OF COMPETENCY

If a judgment on competency is the watershed line for exercising autonomy across a wide range of decision areas, one would expect the concept of competency to be defined and applied without difficulty, except, perhaps, in a few borderline cases. The continental divide is a precise, identifiable line. No doubt there are spots high above the timberline where one can overturn a pail of water and not know its course to the sea, but the direction of any stream is reasonably clear, and the principles for ascertaining the divide's location are easy to state.

The concept of competency at a very abstract level may have such certainty, but its application to many cases in practice does not. Commentators wring their hands over the "elusive definition of mental competency" ([24], p. 53), and despair of formulating precise operational rules for physicians and others facing competency judgments. Roth and Applebaum describe seven different tests that have been used to determine competency, ranging in looseness from ability to evidence a choice to making what in the physician's view is the correct choice [53]. Similarly, Miller analyzes four different standards of competency, from merely making a choice to engaging in certain deliberations in arriving at a choice ([42], pp. 22–28).

The matter of defining competency is actually less difficult than these accounts suggest, once tests for applying the definition, and the substantive outcome that one may be seeking are separated out. Most philosophers and judges who have written about the definition are in agreement that it is the capacity or ability to perform the task of exercising autonomy. As Beauchamp puts it, "the meaning of 'competence' is the ability to perform a specific task", in this case, making decisions or exercising autonomy over medical care.[7] In judicial terms, competency to decide is the ability to understand the nature and consequences of a proposed course of action.

While the philosophers can perhaps sharpen the definition and root out hidden or latent ambiguities, the basic idea is relatively clear. The problem arises when one tries to decide what counts as evidence or proof of this ability. Obviously, one must have some cognitive ability, and the ability to choose. Less clear is how much deliberation one must undertake, and the role which the outcome or results chosen should have in illuminating the underlying ability to choose.

A consensus has emerged that competency to make medical decisions may coexist with civil commitment, fear of the consequences, intermittent lucidity or disorientation, and even irrational behavior. In *Lane v. Candura* (1978) [16], *Quackenbush* (1978) [11], and *Yetter* (1973) [14], the leading decisions on incompetency, the patients whose competency was challenged were intermittently delusional and disoriented, and in the opinion of their doctors, making a misguided or even irrational choice. Indeed, Mrs. Yetter, a 60-year-old woman, had been committed to a mental institution and had delusion about the effect that a mastectomy for breast cancer would have on her hoped-for Hollywood career. Yet in each case the patient was found competent to refuse a life-prolonging procedure, for each patient understood the nature and consequences of the choice presented.

A hard case for courts, commentators, and physicians is how to judge false beliefs or denial of what the doctors see as the reality of the situation. In *Department of Human Services v. Northern* (1978) [7], a 72-year-old woman was found incompetent to refuse amputation of gangrenous feet, even though otherwise lucid and of sound mind, because she refused to believe that she had gangrene and was in danger of dying. As the Tennessee appellate court upholding a finding of incompetency put it:

on the subjects of death and amputation of her feet her comprehension is blocked, blinded or dimmed to the extent that she is incapable of recognizing facts which would be obvious to a person of normal perception.

. . .

For example, in the presence of this Court, the patient looked at her feet and refused to recognize the obvious fact that the flesh was dead, black, shriveled, rotting and stinking.

. . .

The record also discloses that the patient refuses to consider the eventuality of death which is or ought to be obvious in the face of such dire bodily deterioration ([7], p. 197).

A much more defensible position, as Culver, Ferrell, and Green [33] articulated,[8] is to recognize that a person may competently exercise choice although denying certain aspects of a situation. To protect autonomy, they argue for a strict cognitive test of competency that would encompass denial. As long as the patient knows that the doctors think X or Y, the patient should be found competent, even if the patient disbelieves or disagrees with the doctors, or refuses to accept the correctness of their views. Ironically, Appelbaum and Roth [28], in their study of treatment refusals, find that few patients who refuse treatment deny the seriousness of their condition, while the doctors treating them often deny the patient's reactions.

The *Northern* case is indeed open to criticism on this ground. Although purporting to apply the 'appreciate the nature and consequences of the treatment' test, it overlooked evidence that Mary Northern did understand that the doctors had a different view of the matter, and adopted an unduly narrow test of competency. The court even acknowledged inconsistency in admitting that she was competent to consent to the amputation, but not to refuse it, without explaining this discrepancy. The *Northern* case shows the difficulties that can arise when analytic clarity about competency is absent, reactive hostility is generated in the decision-makers, an apparent emergency exists, and the patient is hostile to those intervening in her life ([7], p. 56).

Northern also suggests that the alleged binary nature of the competency judgment is an illusion. The tests for applying the concept to individuals seem confused and often inconsistent, at least in practice. In addition, one suspects that an underlying substantive value position about outcome — about how autonomy should be exercised — influences, if not controls, the assessment of competency. Thus *Lane* (1978) [16], *Quackenbush* (1978) [11], and *Yetter* (1973) [14] may reflect more a view about the value of continued living in debilitated or institutional conditions, than about the concept of competency. *Northern* may be similarly explained, with the Tennessee judges viewing life in her condition as more of a good for a person than the patient or other judges would think. If competency assessments really reflect underlying values about the worth of life and autonomy, and thus represent substantive positions on consequences, the geography of competency must include the consequences that follow the competency assessment. Time spent debating the meaning of competency may be more profitably spent analyzing the consequences that follow such a determination.

IV. CONSEQUENCES OF COMPETENCY/INCOMPETENCY DETERMINATIONS

The watershed nature of the competency determination is to many persons evident in the consequences that follow a finding of competency or incompetency. If competency certifies, as Beauchamp says, that a person is entitled to a significant measure of control over decisions about his or her treatment, then it is this judgment which determines whether the patient or someone else will control whether medical treatments are provided or withheld from the patient. My quarrel is with the automaticity of consequences ascribed to the judgment of competency. Unlike the continental divide, the competency determination does not automatically determine the treatment consequences as a discussion of the consequences of a finding of competency or incompetency will show.

A. Competency

At the outset we must recognize that a finding of competency is relevant to a person obtaining as well as refusing treatment. A judgment of competency may allow a patient to receive an experimental therapy or a major organ transplant, but a competent consent to such a procedure does not obligate others to provide it or create an entitlement to it. Desire, even informed desire, does not create a duty in others to provide everything desired. A judgment of competency at best creates a negative right against treatment, and not a positive right to be treated, unless he has entered into a relationship that entails such duties and has the funds or entitlement to it. As Elizabeth Bouvia[9] has learned, a desire to have a hospital and doctors provide her with a bed and supportive services while she starves herself to death does not create an entitlement to such services.[10] Only if the physician and patient are willing and have the resources to perform a procedure, may competency create a right to treatment.[11]

Even with treatment refusal, the judgment of competency is not nearly as automatic in its consequences as commonly believed. True, a strong presumption in favor of the competent person's right to refuse exists, but no right, including the competent patient's right to refuse medical care, is absolute.[12] A wide variety of circumstances may justify overriding this right. A prison inmate, for example, may not refuse dialysis or even food, no matter how competent, because of the overriding needs of justice, retribution, or sound administration of the prison system.[13] Persons in extreme circumstances, although competent under relevant tests, may be kept alive or treated against their will at least for a period of time to ascertain the stability or certainty of their choice. Even staunch defenders of autonomy, such as George Annas

([27], pp. 507–508), accept treatment of David G., the famous burn patient who appeared to be competently refusing treatment, for a period.

There is also a realm of justifiable paternalism, which would allow competent refusals to be overridden in some circumstances. Culver and Gert [32] articulate a very restrictive test for paternalistic interventions, but nonetheless allow such interventions when their conditions are met. Presumably, then, a competent Jehovah's Witness can be given blood against his will, as long as he will not suffer everlasting damnation, for the benefit to him far outweighs the harm. Similarly, an anorexic near death through a competent refusal to eat may be force-fed [36]. In addition, *Quinlan* (1975) [12] and *Saikewicz* (1977) [3] recognize that preservation of life and the interests of children or other third parties may justify overriding the right. Thus a judgment of competency does not automatically mean that a refusal must be honored. Instead, it leads to an inquiry into whether the paternalistic and nonpaternalistic grounds justifying intervention exist. In many cases they are present, and a treatment refusal, at least for a significant period of time, can be overridden.

The consequences of a judgment of competency are problematic for another reason, which sound medical practice should pay heed to. In many situations, as Jackson and Youngner [39] have shown, a person may be competent and refuse treatment, yet accepting the refusal at face value would be a serious mistake. The patient may be ambivalent or temporarily depressed and change his mind. The patient may be reacting to a transient or situational fear, to a factual misunderstanding, or to conflict with the physician. With more information, explanation, or concern from the physician or others, the patient may change his mind and accept treatment.

Indeed, the danger may no longer be the reluctance but the alacrity with which physicians now honor the patient's refusal, leading to injury or death, avoidable if more time had been spent investigating the source of the patient's concern. The alacrity may stem from a superficial sense of legal and ethical duty. More likely, it is tied to conflict with or hostility to problem patients, or a wish to deny or avoid unpleasant, difficult and demanding patients. Appelbaum and Roth ([28], pp. 1300–1301), in their study of patient refusals, remarked on how easily physicians accepted refusals and how infrequently they made attempts to explore the situation and ascertain the underlying motivation. They ascribe this tendency to physician insensitivity or hostility to the patient, failure to communicate, or denial of the situation.

It is worth noting that too ready an acceptance of a patient refusal could constitute abandonment of a patient, and lead to civil or even criminal liability in extreme cases. For our immediate purposes, however, it reminds us that the judgment of competency again does not and should not lead to an automatic acceptance of the expressed preference for refusal. It merely shifts the inquiry to whether the patient authentically and persistently refuses, and then to when circumstances that justify overriding the refusal are present.

B. *Judgment of Incompetency*

Nor does a valid judgment of incompetency automatically mean that the patient will be treated. When the question of what then to do for the incompetent patient is examined more closely, the putative binary nature of competency again dissolves, with treatment provided or withheld depending on a complex of factors, only one of which is the patient's incompetency.

This conclusion, although often overlooked, should not be surprising. A finding of incompetency certifies only that a patient's expressed preference or lack thereof need not be followed simply because it is voiced. The judgment of incompetency does not tell us what then ought to be done independent of the incompetent patient's speaking or failing to speak on the subject. Indeed, it merely forces us to inquire into how incompetent patients in general and the particular one before us are to be treated, a topic of current legal, ethical, and policy debate.

The answer to this question, however, is neither clear nor obvious. One position likely to be voiced with increasing frequency as the elderly population grows, medical technologies proliferate, and health costs increase, is that a finding of incompetency (at least if it is permanent) should lead to the withholding rather than provision of expensive medical technologies.[14] As the cost squeeze tightens and choices have to be made, permanently incompetent patients could be the last to receive scarce resources because of a judgment that the resources are of less value to them than to competent patients, or a judgment that their lives, being devoid of rationality and the capacity for choice, are of less value.

Until recently the courts, policy-makers, and others have avoided pronouncing stark judgments on the relative worth of incompetent persons. Yet such judgments have always been implicitly made, if only in the manipulation of purportedly patient-centered tests such as substituted judgment to reach the result of nontreatment [49]. *Quinlan* (1975) [12] and *Spring* (1980) [13], important cases in defining the criteria for withholding treatment from incompetent patients, do not withstand scrutiny in their own terms of substituted judgment, yet make sense as decisions to conserve some resources for the sake of others that could benefit incompetent persons. As the cost issue moves from the closet and is discussed openly, allocation and rationing criteria will probably emerge that exclude some incompetent patients from expensive technologies. A judgment of incompetency would then mean no treatment, rather than the provision of treatment, as is commonly thought.

More likely is continuation of the currently dominant patient-centered approach to decisions about incompetent patients. Rather than focusing on resource constraints and the needs and interests of family, taxpayers and other patients, a patient-centered approach strives to show respect by doing what will benefit or serve the needs of the incompetent person. But a strict patient-centered approach will not automatically yield a decision to treat in all cases of incompetency. Whether treatment is required to respect the incompetent patient's interests will depend on the test chosen for respecting the person of the incompetent. If we adopt a best interest test (in my view the only defensible approach) [49], the decision to treat or not will be made in terms of what course will best serve the interest and needs of the incompetent patient, in light of his current condition and prognosis. (On this view past wishes and preferences about what he would now choose are not relevant, unless he will regain competency.)

But on this test, it will often be the case that despite incompetency, further treatment will not be in the patient's interests, and thus should not be provided. Amputation of the leg of Mary Northern or Quackenbush is arguably not in their interests, even if they were incompetent, because continued life, in light of the painful adjustments and change in living required, is, from their own perspective, too much of a burden. True, the operation will keep them alive, but continued life at the high price entailed may not be in their interest, anymore than it would be in Saikewicz's interest to have his life prolonged a few months through very painful chemotherapy. Indeed, their expressed preferences against the operation, even though found incompetent, should be given weight here as an indicator of the burdens to them of the proposed treatment — weight that may show that it is not in their interest to have the treatment in light of their condition and prognosis. A finding of incompetency thus just pushes the inquiry to another level, where an outcome of always treat is very problematic.

If one adopts a substituted judgment or prior directive test for respecting the incompetent patient's interests under the patient-centered approach, an outcome of treatment upon a judgment of incompetency is even less likely. One finds in the President Commission's Report [47] and throughout the literature a claim that we are obligated to do what the patient, when previously competent, directed should be done if he ever becomes incompetent. In its most narrow form, this obligation arises only if the patient has made a clear prior directive, either written in the form of a living will or orally to others, of how he wants to be treated if ever incompetent. In its broader form it is said that we are obligated to do what we would infer, in light of his past preferences, values, disposition, and character, the person would now want done, even though he has not specifically addressed this situation and cannot now speak to us.

Now I believe that the narrow or broader position of prior direction (substituted judgment in the parlance of some) is seriously mistaken as an approach to showing respect for the incompetent patient before us [49]. The present needs and interests of the incompetent patient cannot be accurately assessed in terms of what he had previously chosen when competent or would choose if competent in light of his past preferences, because he no longer has those preferences. Unless the incompetency is transient, to treat him as the competent person he was is to treat him as a very different person, for he is now incompetent and has different needs and interests. A policy of adhering to prior directives made while competent may sacrifice the needs and interests of the incompetent patient to the satisfaction that people derive from knowing that they can control their lives in advance, and from the avoidance of discomfort entailed in having to face directly the needs of incompetent patients.

However, I need not pursue this argument, for if one adopts this position, then prior directives against treatment would have to be honored when a patient becomes incompetent. Incompetency would invariably mean not treating, rather than treating, as is commonly assumed in discussing competency, at least where the person has previously spoken against treatment or such a choice would be inferred, if he were treated as if he retained his previous values and preferences. This result is clearest when the person competently refuses treatment, and then as a result of the withheld treatment lapses into unconsciousness. If prior directives control, the mere fact that the patient is now incompetent should not lead to treatment. (Nor would a best interests test necessarily lead to treatment in such cases.)

Thus the prior directive approach would, where evidence of a directive or expressed wish is present, allow the prior wishes against treatment to control upon a finding of incompetency, and not the physician's automatic assumption that if incompetent, treatment is justified. An automatic assumption of treatment would not even follow when a prior directive is absent or the inferred wishes are unascertainable. At that point the best interests test would control. As we have seen, it is problematic whether treatment will be in the interests of an incompetent patient. Once again the continental divide of competency vanishes. Determining competency or incompetency will not in itself prescribe the course of action. The patient's mental state becomes one of many relevant factors in deciding how to proceed, but in itself does not tell us whether the waters of treatment will flow toward or away from the patient.

V. PROCEDURES FOR THE COMPETENCY JUDGMENT AND ITS AFTERMATH

A guide to the geography of competency should also address the procedures for making these judgments. If the patient is retarded, unconscious, or otherwise clearly incompetent, the question moves directly to the next stage of determining what the patient's interests or the legitimate interests of others demands. In cases where incompetency is not so clear, the procedural fulcrum is of course the physician (or nurse) who must identify competency as an issue to be addressed.

As widely noted, physicians are much more likely to address the competency issue when a patient refuses than when a patient consents to a recommended course of therapy. As graphically demonstrated in *Lane v. Candura* (1978) [16], no one questioned Ms. Candura's competency when she consented to the two previous amputations. It was only when she rejected the operation that any question of her competency arose.[15] Similarly, a looser test of competency may apply, depending on whether the patient is going along with the doctor or refusing (though we now see a trend toward applying a loose test to allow refusals). Knight points out how the risk/benefit ratio of the proposed treatment is relevant to the physician's assessment of competency:

... the risk/benefit ratio is a major factor in determining the competency of a particular patient in accepting or refusing treatment. If, for a serious illness, risk in treatment is minimal for a serious illness and benefit great, the patient who refuses treatment will be declared incompetent. If risk is great and benefit is minimal, the patient who refuses is likely to be declared competent ([40], p. 3).

Although competency — capacity for choice — should not vary in this way, one can understand this tendency in physicians. If the patient is doing what the physician recommends, the patient would seem to be exercising good sense and hence be competent. Moreover, a treatment intended to sustain life is less likely to lead to lawsuits and other complications than is accepting a patient's refusal. Then, too, standing by as the patient worsens when treatment is available may stimulate the physician's own fears about death and doubts about his professional abilities, which might explain the physician's reaction to the refusal of therapy for throat cancer in the case described by Knight ([40], pp. 8—9), and the zeal of the would-be rescuers in *Northern* [7].

A procedural account of the competency judgment is no more clear or binary than the other aspects of competency that we have examined. The relevant issues and considerations about procedures for competency judgments depend largely on the stage of inquiry — is the doctor deciding to inquire into competency, evaluating competency, or deciding on what to do in light of the assessment of competency made?

For example, the physician's tendency to avoid issues of competency when the patient consents is generally not problematic. Yet there are situations in which the physician should be slow to assume that acceptance of the doctor's recommendation indicates competency and the lack of legal and ethical concern. The physician, for example, should not readily assume competency if the procedure is highly risky. The competency of Irwin Karp or Barney Clark to consent to experimentation with a heart prosthesis should not be presumed merely because they agree with the surgeon's willingness to experiment. Nor should the borderline retarded woman's consent to the doctor's sterilization, or the

14-year-old's acquiescence on the bone marrow transplant be found to be competent without more scrutiny. Indeed, in such cases the special protective procedures for determining competency in the mental health context should be followed as well ([46], [15], [1], [52], [48]). (Hospital policies for inquiring further into competency in certain categories of problematic cases may be desirable.)

If the physician does pause over competency, as should occur when the patient refuses treatment, the procedural issue then becomes whether the physician should make the competency decision alone, call in a psychiatric consultant, a hospital ethics committee, or even go to court. In most cases the physician pausing over competency will end up making the judgment alone, rarely even calling in a psychiatric consultant,[16] and even more rarely seeking the advice of a hospital ethics committee or court. They key issue thus is whether the physician can adequately make this assessment, or whether consultation or review is necessary.

In most cases the physician may be adequately equipped to assess competency. If the standard of competency is articulated in hospital policy, and a checklist of mental function (such as the one proposed by Annas ([27], p. 511)) is followed, the physician should be able to determine competency alone. Autonomy and discretion in making professional judgments is highly valued by society (and the profession), and there is no reason why the professional judgment about competency should be reviewed or limited procedurally here anymore than are other clinical judgments. The incentives to use reasonable care provided by the malpractice system should be sufficient protection here as well.[17]

But good clinical practice recognizes the need for consultation, and so there will be times when review or consultation on the physician's competency judgment is in order. However, unless the competency is truly binary as to consequences, which as we have seen it seldom is, the review of competency will intrude into the question of what then to do in light of the patient's mental capacity to express preferences. If the competency assessment is truly key, in some cases a psychiatric consultation will be sufficient to show reasonable care.[18] In more difficult cases, especially where the standards of competency are less certain, e.g., does denial of medical facts make a person incompetent, review by a hospital ethics committee may be required by the duty to use reasonable care in ascertaining and applying the standard. The consent committee used in *Kaimowitz* (1973) [15] to review the competency and validity of consent to experimental psychosurgery, and occasionally recommended for enhancing the consent process in experimentation with special populations ([1], [52]), may be desirable in some non-research settings [38]. In case of disputes or very serious cases, such as sterilization of the retarded, judicial review may also be desirable to assure advance immunity, even though the failure to go to court should not in itself be penalized, if the underlying decision were reasonably made, as through consultation with a hospital ethics committee.

Perhaps the most important procedural issues in the domain of competency arise over what then to do in light of the assessment made, rather than over the assessment itself. Indeed, as we have seen, the two often merge and are difficult to untangle. If the competency decision really cloaks an underlying substantive value choice of what to do, or if the competency judgment leads us there anyway, a procedural issue will then be what procedures, if any, to follow in determining what to do in light of the competency assessment. Here again, the procedures will vary along several lines.

If the patient is found competent, the question of going along with wishes for treatment should not be problematic, if the procedure has been established as safe and effective, or is part of a research protocol. On the other hand, if the patient refuses, the physician should not treat over the patient's wishes without some review of the justification of that course of action. If autonomy is a key value, the physician alone or society should not override it without some impartial proceeding that establishes its necessity, such as a court or a hospital ethics committee.[19]

Less obvious, however, is the physician's need for review of a decision to go along with the patient's refusal, where death or serious harm will result. Not only should the physician heed the insights of Jackson and Youngner and probe more deeply into the patient's thinking and feeling,[20] but the physician may also gain from consulting others. In extreme cases, such as Bouvia,[21] resort to the courts may be essential to ascertain the proper course of action. In most cases, however, a hospital ethics committee could provide the necessary consultation.

If the patient is found incompetent, procedural protections may also be in order. Again, the need for them may seem more obvious in the case of withholding a life-prolonging treatment, since the patient appears more obviously injured by such a decision. The question here is the policy question of what procedures to follow in withholding treatment from incompetent patients, currently a major public policy issue. While a presumption in favor of allowing the physician and family to decide these questions has been recognized [47], there is a recognition that review of their decision and even appointment of a surrogate decision-maker is in order in some cases, such as handicapped newborns. Again, a hospital ethics committee can play a leading role here, with resort to the courts only in case of disputes or significant cases.

Providing treatment to the incompetent patient does not appear procedurally problematic. Yet there are patient interests in avoiding excessive treatment, and social interests in conserving resources that may justify some review of the physician and guardian's decision. Utilization review committee and PSROs performed such a function. Hospital ethics committees or similar bodies may also take it on.

VI. CONCLUSION

I have tried to provide a map of the geography of competency which shows that its apparent binariness or significance as a decision fork is more problematic than is generally thought. Competency issues arise across the landscape of medical ethics with children, the mentally ill, and adults over decisions from experimentation and sterilization to passive suicide through treatment refusals.

Given the high value accorded autonomy and self-determination in our society, a focus on competency as watershed seems natural. Yet the focus soon blurs, because of difficulties in applying the concept. As the focus blurs, it also shifts to the substantive and procedural questions surrounding the consequences of a competency or incompetency finding.

With this shift in focus, the person's competency still is relevant, but is not necessarily determinative. Competency leaves the foreground and merges into the complex of considerations that enter into judgment of what to do. The patient's mental awareness and capacity to exercise choice is a relevant and important factor in making the decisions presented. One cannot easily show respect for the patient without taking his capacity for choice into account, for respect will often entail respecting his choices. As one factor (a particularly weighty one) among a complex of relevant factors, however, the judgment about the patient's capacity will not be as determinative as is commonly assumed.[22] A judgment of incompetency, for example, merely opens up a new set of choices and considerations to take into account.

What then are we to do about competency? Perhaps it is time to change the metaphor. Let us stop talking about competency as a continental divide of treat or do not treat. Although not a clear watershed, it is a useful but not definitive map as to where the continental divide is. Competency shows us the path or the general direction to the solution, but it is not the path itself.

Rather, the competency finding is a map of considerations that are relevant. Competency is a filter or screen that channels our thinking by limiting the alternatives and factors to be considered. Like a command function of a computer, it opens up new issues, although unlike a computer, it is not totally neutral about the decisions and questions that follow. The final decision can be reached only after leaving the narrow domain of competency and confronting the value choices between a patient-centered or other-directed approach that are presented.

I find the interesting questions about competency to be the relationship between substantive results and the competency label. The relationship is usually thought of as distinct, a causal link between two separate things. As we have seen, however, the connection is often much closer and the separation weaker than thought. Indeed, there is a powerful tendency for judgments about results to infiltrate and colonize the competency decision, often when clear thinking about the value-choices embedded in the results is most needed. I suspect that the main function of concepts such as competency (substituted judgment, living wills, choice of surrogate are others) may be to hide the trade-off between a patient-centered and other-directed approach that is inevitable, but painfully acute, in the medical endeavor.[23] Since we want to have both, and cannot, we prefer to avoid the issue by pretending we are playing a different game.

The concept of competency sometimes confuses the real game being played. But properly understood and placed in context, it may help us to play the real game better. Used as a rough map of difficult terrain, it may point us in the direction of the path that will lead us to decisions that we can live with.

University of Texas Law School
Austin, Texas

NOTES

[1] "The root premise is the concept, fundamental in the American jurisprudence system, that 'every human being of adult years and sound mind has a right to determine what shall be done with his own body'" (*Canterbury v. Spence* (1972) ([5], p. 781), quoting Justice Cardoza in *Schloendorff v. Society Hospital* (1914) [21].

[2] The relation between competency and informed consent is discussed in Beauchamp [29].

[3] For an account of the limits of reproductive autonomy, see Robertson [50].

[4] The question of what respect for persons entails is especially difficult in the case of incompetent persons.

[5] Exceptions to the requirement of informed consent of research subjects exist where the patient is a child, retarded or an institutionalized mentally ill person, and the research is for the person's benefit or imposes minimal risk. See National Commission for the Protection of Human Subjects of Biomedical and Behavioral Research ([34], [35]).

[6] A good account of the deficiencies in the consent process in this case is found in G. Annas [26].

[7] Thus Beauchamp's distinction between autonomy and competence ([29], pp. 59—62), while relevant to any general exposition of the subject matter, does not apply when the performance competency at issue is the exercise of autonomy itself. Indeed, he recognizes that "being an autonomous person may be *the* solely sufficient criterion that we use in our culture for general (but not necessarily specific) competence" ([29], p. 60).

[8] Presumably an emaciated woman suffering from anorexia nervosa who understands that the physician is telling her that she is undernourished almost to the point of death, but who continues to believe that she is grossly overweight, would under this approach be found competent. Of course, a finding of competency would not automatically mean acquiescence to her refusal to eat. See Dresser [36].

[9] A useful summary of this case is provided by Annas [25].

[10] The physician who has already entered into the relationship with the patient may have an obligation to provide the treatment or withdraw from the relationship. In some cases there may also be an obligation to provide supportive care, as when it is not possible to transfer the patient to another doctor's care and the patient will suffer if the physician withdraws. See Robertson [41].

[11] The right claimed here is still a negative and not a positive right, for it is a claim against the state to refrain from interfering in doctor-patient relations, and not a claim against the state or doctor to provide physician services. The right of a willing doctor and patient to undertake a medical procedure for the patient's benefit without state interference is arguably protected by the constitutional right of privacy that protects refusal of treatment. The argument for this claim can be found in *Andrews v. Ballard* (1980) [2] and in Chief Justice Bird's dissent in *People v. Privatera* (1979) [17].

[12] Presumably even the strong statement of the right to refuse treatment by the Legal Advisors to Concern for Dying in their Model Treatment Act, which does not specify any exceptions, would permit an exception in some of the cases discussed below ([31], p. 921).

[13] See, e.g., *Commissioner of Corrections v. Myers* (1979) [6], *State ex. rel. White v. Narick* (1982) [22].

[14] For a general description of the problem, see Aaron and Schwartz [23].

[15] The same phenomenon was explicitly noted in the Northern case. See Abernethy ([24], p. 55).

[16] Applebaum and Roth note that psychiatrists were consulted only seven of 105 patients who had refused treatment. All seven cases involved a patient who had had a past psychiatric history or was overtly depressed or psychotic ([28], p. 1300).

[17] Malpractice litigation, despite the fears and misconceptions of doctors, is rare relative to actual incidence of malpractice; and therefore may not be an effective constraint except in extreme cases. See Schwartz and Komesar [54].

[18] The psychiatric consultant may have a conflicting role and be used by physicians to bring about a course of conduct thought desirable by the physician. See Perl and Shelp [36].

[19] The patient, of course, should be adequately represented and the requirements of procedural due process satisfied in any decision process.

[20] In his analysis of a severely burned patient's refusal of anti-infection therapy, *Taking Care of Strangers*, Robert Burt [30] also shows how a refusal of treatment may mask other concerns that need to be uncovered if the patient's autonomy is to be respected.

[21] The Bouvia case was perhaps appropriately brought to court. Other cases have not been, as the diabetes case in New York shows.

[22] Competency could, of course, be determinative if one specified the context and circumstances narrowly enough, e.g., a competent adult's refusal of a very risky operation when she has a metastatic cancer.

[23] G. Calabrese and P. Bobbitt's *Tragic Choices* [31] is the classic exposition of this tendency.

BIBLIOGRAPHY

Cases
1. Aden v. Youngner, 57 Cal. App. 3d 662 (Cal. Ct. of Appeals, 1976).
2. Andrews v. Ballard, 498 F. Supp. 1038 (S.D. Texas, 1980).
3. Belcherton v. Saikewicz, 373 Mass. 728, 370 N.E. 2d 417 (1977).
4. Bellotti v. Baird, 443 U.S. 622 (1979).
5. Canterbury v. Spence, 464 F. 2d 772 (D.C. Cir. 1972).
6. Commission of Corrections v. Myers, 399 N.E. 2d 452 (Mass. 1979).
7. Department of Human Services v. Northern, 563 S.W. 2d 197 (Tenn. Ct. of Appeals, 1978).
8. In re Grady, 85 N.J. 235, 426 A. 2d 467 (1981).
9. In re Guardianship of Eberhardy, 102 Wis. 2d 539, 307 N.W. 2d 881 (1981).
10. In re Mary Moe, 385 Mass. 555, 432 N.E. 2d 712 (1982).
11. In re Quackenbush, 156 N.J. Super. 282, 383 A. 2d 785 (1978).
12. In re Quinlan, 70 N.J. 10, 355 A. 2d 647 (1975).
13. In re Spring, 405 N.E. 2d 115 (Mass., 1980).
14. In re Yetter, 62 PA. D & C 2d 619 (1973).
15. Kaimowitz v. Dept. of Mental Health, Cir. Ct., Wayne County, Mich., July 10, 1973.
16. Lane v. Candura, 376 N.E. 2d 1232 (Mass. App. 1978).

17. People v. Privatera, 23 Cal. 3d 697, 153 Cal. Rptr. 431, 591 P. 2d 919 (1979).
18. Rennie v. Klein, 462 F. Supp. 1131 (D.N.J. 1978).
19. Rogers v. Okin, 478 Supp. 1342 (D. Mass. 1979), aff'd in part, rev'd in part, vacated and remanded, 634 F. 2d 650 (1st Cir., 1980), cert. granted April 20, 1980.
20. Satz v. Perlmutter, 363 So. 2d 160 (1978).
21. Schloendorff v. Society Hospital, 211 N.Y. 125, 105 N.E. 92 (1914).
22. State ex rel. White v. Narick, 292 S.e. 2d 54 (W. Va. 1982).

Articles and Books
23. Aaron, H. and Schwartz, W.: 1984, *The Painful Prescription: Rationing Hospital Care*, Brookings Institute, Washington, D.C.
24. Abernethy, V.: 1984, 'Compassion, Control, and Decision About Competency', *American Journal of Psychiatry* **140**, 53—58.
25. Annas, G.: 1984, 'Case of Elizabeth Bouvia: When Suicide Prevention Becomes Brutality', *The Hastings Center Report* **14**, 16.
26. Annas, G.: 1983, 'The Lion and the Crocodile', *The Hastings Center Report* **13**, 20—22.
27. Annas, G. and Densberger, J.: 1984, 'Competency to Refuse Medical Treatment: Autonomy v. Paternalism', *Toledo Law Review* **15**, 561—596.
28. Appelbaum, P. and Roth, L.: 1983, 'Patients Who Refuse Treatment in Medical Hospitals', *The Journal of the American Medical Association* **41**, 1296—1301.
29. Beauchamp, T. L.: 1991, 'Competence', in this volume, pp. 49—77.
30. Burt, R.: 1979, *Taking Care of Strangers*, The Free Press, New York.
31. Calabresi, G. and Bobbitt, P.: 1978, *Tragic Choices*, Norton and Company, New York.
32. Culver, C. and Gert, B.: 1979, 'The Justification of Paternalism', *Ethics* **89**, 199—205.
33. Culver, C., Ferrell, R. B. and Green, R. M.: 1980, 'ECT and Special Problems of Informed Consent', *American Journal of Psychiatry* **137**, 586—591.
34. 'Death of Boy-in-Bubble as Amazing as Life', *New York Times*, February 24, 1984, p. 1.
35. Doudera, E. and Swazey, J. (eds.): 1982, *Refusing Treatment in Mental Health Institutions — Values in Conflict*, Health Administration Press, Ann Arbor, Michigan, pp. 19—30.
36. Dresser, R.: 1984, 'Feeding the Hungry Artist: Legal Issues in Treating Anorexia Nervosa', *Wisconsin Law Review*, 297—374.
37. '15 Days Out of Bubble, David Dies', *The Houston Post*, February 23, 1984.
38. 'Girl Gets Cancer Treatment Despite Father's Objections', *New York Times* September 22, 1983, p. 8.
39. Jackson, D. L. and Youngner, S.: 1979, 'Patient Autonomy and Death with Dignity: Some Clinical Caveats', *New England Journal of Medicine* **301**, 404—408.
40. Knight, J.: 1991, 'Judging Competence; When the Psychiatrist Need, or Need Not, Be Involved', in this volume, pp. 3—28.
41. Legal Advisors Committee, Concern for Dying: 1983, 'The Right to Refuse Treatment: A Model Act', *American Journal of Public Health* **73**, 918—921.

42. Miller, B.: 1981, 'Autonomy and the Refusal of Lifesaving Treatment', *The Hastings Center Report* **11**, 22—28.
43. National Commission for the Protection of Human Subjects of Biomedical and Behavioral Research: 1978, *The Belmont Report*, U.S. Government Printing Office, Washington, D.C.
44. National Commission for the Protection of Human Subjects of Biomedical and Behavioral Research: 1978, *Research Involving Children*, U.S. Government Printing Office, Washington, D.C.
45. National Commission for the Protection of Human Subjects of Biomedical and Behavioral Research: 1978, *Research Involving Those Institutionalized as Mentally Infirm*, U.S. Government Printing Office, Washington, D.C.
46. Perl, M. and Shelp, E. E.: 1982, 'Psychiatric Consultation Masking Moral Dilemmas in Medicine', *New England Journal of Medicine* **397**, 618—621.
47. President's Commission for the Study of Ethical Problems in Medicine and Biomedical and Behavioral Research: 1983, *Deciding to Forego Life-Sustaining Treatment*, U.S. Government Printing Office, Washington, D.C.
48. Robertson, J. A.: 1984, 'Ethics Committees in Hospitals: Alternative Structure and Responsibilities', *Quality Review Bulletin* **10**, 6—10.
49. Robertson, J. A.: 1987, 'Equal Respect and Distributive Justice in Nontreatment Decisions for Incompetent Patients: A Comment on Buchanan', in D. Regan and D. Van deVeer (ed.), *Border Crossing*.
50. Robertson, J. A.: 1983, 'Procreative Liberty and the Control of Conception, Pregnancy, and Childbirth', *Virginia Law Review* **69**, 405—464.
51. Robertson, J. A. : 1982, 'Taking Consent Seriously: IRB Interventions in the Consent Process', *IRB: A Review of Human Subjects' Research* **4**, 1—5.
53. Roth L. H. and Appelbaum, P.: 1981, 'Clinical Issues in the Assessment of Competency', *American Journal of Psychiatry* **138**, 1462—1467.
54. Schwartz, W. and Komesar, K.: 1978, 'Doctors, Deterrence, and Damages', *New England Journal of Medicine* **29**, 1282—1289.

PATRICIA D. WHITE AND SUSAN HANKIN DENISE

MEDICAL TREATMENT DECISIONS AND COMPETENCY IN THE EYES OF THE LAW: A BRIEF SURVEY[1]

The question of mental competence arises in many different areas of the law. In order to make contracts and wills, have criminal responsibility, be a witness, or stand trial, a person must be mentally competent. Obviously, however, not every degree of mental deficiency or peculiarity is legally significant. Moreover, there is no single test of competence for these various contexts; rather, competency is defined in different ways for different purposes [56]. As a result, and somewhat paradoxically, a person can be legally incompetent in some areas while remaining legally competent in others. For example, since the law requires a lower degree of capacity in the making of wills than it does for the execution of contracts, one may be competent to make a will but incompetent to transact ordinary business affairs.

The law has only recently begun to be concerned with determining the competency of patients to make decisions concerning their medical treatment.[2] The tests used by courts to determine competency in this new context mirror those used to determine the capacity to contract [38]. This has likely occurred because, except in certain emergencies, the law does not allow nonconsensual medical treatment. (See, e.g., [4], [8], [23], [41], [51], [53]) A doctor must, in certain respects, enter into a contract with a patient in order to diagnose and treat him ([36], [65]). In order to explore the use of legal competency as a condition of making treatment decisions, we must therefore trace the development of the notion of competency in contract law.

I. COMPETENCY TO CONTRACT

It has long been a principle of the common law that contractual obligations can be assumed only by free and intelligent consent ([48], [42]). Traditionally, therefore, an incompetent person who was incapable of giving such consent could not form a valid contract ([42], [46]). This view prevailed in the United States well into the nineteenth century. An incompetent person, having "nothing which the law recognizes as a mind", could not in principle make a contract, for this required the assent of two minds ([7], [42]). He simply did not have mind enough for the "meeting of the minds" required to contract. As a result, the mental incompetency of one of the parties made a contract entirely void ([30]; [43], p. 496).

A more recent view has been that incompetents require protection from their own actions.[3] Thus, rather than being void, their contracts are 'voidable' ([43], p. 497). The contract is voidable at the election of the incompetent or, more likely, his personal representative, but recission will be denied if the other party to the contract was, at the time of the contact, ignorant of the incompetence and the *status quo* cannot be restored. The interests of those who deal with the disabled person and who are unaware of the disability are thereby given some measure of protection [43]. But, also in contrast to earlier law, it is nonetheless possible for the incompetent party to enforce a transaction that has proved advantageous to him.

The test most frequently used to assess a person's competency to contract is what has become known as the 'understanding test' [57]. Although the phraseology varies among and within jurisdictions,[4] the common element is a focus on the person's ability to understand and appreciate the nature and consequences of the act or transaction in which he is engaged [40]. It is crucial that the person understand the nature and effect of the particular act or transaction that is challenged [45]. This helps explain why there are different standards of competency for different legal contexts. The test itself need not vary, just its application. Since some contexts are simpler than others, the capacity needed to understand their nature and consequences differs appropriately.

The understanding test for competency to contract is perhaps better approached by looking at what does *not* constitute incompetency. Proof of mental weakness or even of insanity are not in themselves enough to show incompetency; rather, the disorder must be shown actually to have destroyed the capacity to understand the questioned transaction [3], [47]. Other factors that alone have been held to be insufficient to prove incompetency include *senile dementia*, old age, the influence of drugs, physical distress and pain, and commitment ((([57], p. 1032) and the cases noted there). None of these is sufficient because with each it is sometimes possible for the afflicted person to have 'lucid intervals' during which he does have the requisite capacity to contract (e.g., [3], [33]). The key is twofold. In order to render someone incompetent, a mental disorder must be significant enough to have made the person incapable of having understood the situation and it must in fact have impaired the person's capacity to understand the particular transaction in question.

In addition to the 'understanding test' of incompetency to contract, there is what has been referred to as the 'insane delusion' test. An insane delusion, the belief in the reality of facts that do not exist and in which no rational person would believe, can lead to the avoidance of a contract if the delusion is deemed to be sufficiently related to the contract's subject matter [45]. A delusion will not, however, affect the validity of a contract if it concerns a matter that is not connected with the transaction itself. The question then becomes whether the contract was *motivated* by the delusion [45].[5]

There is some support for the view that the insane delusion test is a distinct and sufficient test of contractual incompetence. The claim is that a person who understands what he is doing but who is motivated by a deluded belief is as much in need of protection as one who cannot understand what he is doing [57]. This rationale uses the term 'understanding' in a limited sense. Alternatively, an insane delusion may be viewed as a species of failure to understand. On this view, the test is merely one of the ways in which the understanding test may be applied ([51], [38]).

A more recent trend in assessing competency to contract focuses on a person's ability to control his actions in relation to the transaction. Thus, "[i]ncompetence to contract also exists when a contract is entered into under the compulsion of a mental disease or disorder but for which the contract would not have been made" [9]. While this view found some support in dicta in the 1950s (see [57]), it was first articulated by a court in 1963 in *Farber v. Sweet Style Manufacturing Corp.* [9]. In this case, a contract made by a manic-depressive was voided where the illness was recognized as affecting his motivation but not his ability to understand. This test now finds support in the *Restatement (Second) of Contracts*. A person is incompetent to contract if, among other things, "he is unable to act in a reasonable manner in relation to the transaction" (Section 15) [37].

The "ability to control one's actions test" is thought to be important because the understanding and insane delusion tests "fail to account for one who by reason of his mental illness is unable to control his conduct even though his cognitive ability seems unimpaired" [26].

For a long time, the only competency to contract tests articulated by courts were versions of the understanding test or the insane delusion test (however conceived). Milton Green, a leading commentator on mental incompetency in the law, noted, however, that in many cases the articulated standard was being ignored and an unarticulated one was being followed [47]. He found, not surprisingly, that courts, rather than focusing on whether the alleged incompetent could understand, often directed their attention to ascertaining whether the transaction itself turned out to be a fair one. They would test the competency of the person only after finding that it had not. Green felt that this 'inarticulate standard' was the actual rationale for the results in many difficult cases where the understanding test could not be easily applied.

Green's 'inarticulate standard' has since been articulated by a number of courts. For example, in *Ortelere v. Teachers' Retirement Board*, a 1969 New York case, a contract was voided where the choice made by the person judged incompetent was felt to be "so unwise and foolhardy that a factfinder might conclude that it was explainable only as a product of psychosis" [26]. *Krasner v. Berk*, a 1974 Massachusetts case, was also very candid in its use of objective tests of overall fairness [18]. This case was a very close one on sufficiency of understanding; there was evidence that showed *some* understanding of the terms of the agreement. The court stated that in such a case the controlling consideration should be "whether the transaction *in its result* is one which a reasonably competent person might have made"[6] (emphasis supplied). Thus the court based its finding on the actual outcome of the transaction, not on whether the alleged incompetent actually understood its nature and consequences.

In contracts, as well as in other areas, the law starts with a presumption of competency [11]. The burden of proof in competency proceedings is therefore on the party alleging incompetence. In a proceeding for the appointment of a guardian, for example, the burden is on the petitioner to prove that the alleged ward is incompetent [34]. In a case seeking to avoid a contract, the burden of persuasion is on the person who asserts incapacity to contract [57].

The question of legal competency is viewed as a question of fact to be determined from the circumstances of each individual case [10]. This is consistent both with the idea that the competency determination tests an individual's understanding of the nature and likely consequences of a *particular transaction*, and with the idea that it can test the beliefs that motivated him in the transaction. Only by looking at the facts surrounding each individual case can either determination be made. The question is thus ultimately determined by the trier of fact — either a judge or jury. The testimony of psychiatrists will not be decisive, for the central focus on one transaction is not the ordinary one in the course of diagnosis and treatment ([43], p. 503). By focusing on competency as a question of fact, the law also recognizes that a person can be competent in some areas but incompetent in others, or that a person's competency can change from day to day even within one context [38].

Many of the competency issues raised in the contact of contract law carry over into the area of medical care decision-making. Against this backdrop, we shall now explore how the notion has developed there.

II. COMPETENCY TO MAKE MEDICAL CARE DECISIONS

The importance of competency in the medical treatment area derives from the common law notion of consent. An often-quoted passage from a 1914 case sets forth the principle that "[e]very human being of adult years and sound mind has a right to determine what shall be done with his own body" [28]. A physician who treats a patient without consent is viewed as committing a battery — an intentional interference with a person without his or her consent [49]. More recently, the law has gone further and required not only that consent be obtained, but that it be *informed*. To comply with the requirements of informed consent, a physician must give a patient all of the information that might reasonably influence his decision, including information about his condition, the risks and benefits of the proposed treatment, and its alternatives [49].

There are only a few exceptions to the requirement that a physician obtain the patient's informed consent before he begins to treat him. A doctor need not obtain consent in an emergency situation and in unusual circumstances the doctor may withhold information from a patient if he thinks that the information itself would harm the patient [57]. Additionally, and most importantly for our present purposes, the informed consent requirement is excepted when the patient is not competent; an incompetent patient lacks the capacity to give valid consent.

When a patient is not competent, the physician must obtain consent to treat him from someone else. The patient's right to make his own treatment decisions is transferred to another person (or to a court or to an institutional review board [38], and the patient may well be treated against his expressed desires.[7] Thus the competency determination is crucial.

It is not surprising, given the similarities between the patient-physician relationship and a contractual one, that the usual test for determining competency to make treatment decisions has turned out to be the same as that usually used to determine competency to contract — whether the person can understand and appreciate the nature and consequences of his actions [22]. While, as for contracts, this standard has been criticized as being vague [44] and overly general [55],[8] it is given meaning by the specific context in which it is used. Here, once again, the key question is whether the person understands the nature and effect of the *particular* transaction being contemplated. Does the patient understand the treatment decision with which he is faced? Just as the capacity needed to make a simple will is lower than that needed to make a complex contract, so is the capacity standard lower for treatment decisions that are fairly straightforward and involve little risk [61]. Accordingly, the most stringent competency standards are found in the area of experimental treatment [60], where it is the most difficult for patients to understand and appreciate the nature of the treatment decision, and where the risks are often very high relative to the potential benefit [61].

Historically, an incompetent's contracts were void because the incompetent lacked the mental wherewithal required for "a meeting of the minds". Today they seem to be voidable, at least in part because of some sense that the incompetent needs protection. Some variant of the latter motivation must underlie the requirement of patient competence in informed consent. The doctor wants (we hope) to do what he thinks is best for the patient, but he needs the patient's informed consent in order to proceed. If the patient is incompetent and either refuses or is incapable of giving the needed permission, we do not simply deny him medical treatment (thereby, in effect, voiding his relationship with any doctor). Rather, we appoint another to look out for his best interest and allow that person (or court or institutional review board) to substitute his (or its) judgment for the patient's in dealing with the doctor. The requirement of patient competence thus seems designed to attempt to maximize (or at least increase) the chances that the actual treatment decisions made with respect to that patient are in fact in his best interest.

While most of the recent cases which involve competency to make medical treatment decisions focus on an adult patient's capacity to refuse treatment ([38], p. 573), the earlier medical competency cases typically had a rather different posture. They tended to be suits brought by the patient against the doctor and they sought damages for assault and battery for alleged unauthorized treatment. (See, e.g., [10] and [24]).

These cases are highly unusual today, in large measure one suspects, because many were successful. In recent years, hospitals and physicians have become scrupulously careful to establish and follow standardized procedures for obtaining patient consent to treatment and, now, determining competency to consent. In recent times what we more frequently see are doctors questioning the competency of patients to give or withhold that consent.[9] Such cases are usually brought to court through the invocation of state statutory provisions that allow the appointment of a guardian for the purpose of making treatment decisions for one judged mentally incompetent [44].

The discussion of incompetency that is found in these recent cases is instructive. The concerns and standards expressed in them mirror those found earlier in the discussions of incompetency to contract. Probably the most often cited of the cases involving the question of competency to refuse medical treatment are *Lane v. Candura* [19], *In re Quackenbush* [16], *In re Yetter* [17], and *Department of Human Services v. Northern* [6]. It will be useful to look briefly at each of them.

Lane v. Candura [19] involved a 77-year-old woman who refused her doctor's recommendation that her gangrenous leg be amputated. Her daughter petitioned the court to be appointed temporary guardian with authority to consent to her mother's operation. In holding that Mrs. Candura was competent to decide for herself, the appellate court overturned the trial court's determination that Mrs. Candura was incompetent because her decision was irrational. Stressing that the irrationality of her decision did not justify a finding of incompetence in the legal sense, the higher court found Mrs. Candura competent because she was capable of appreciating the nature and consequences of her acts. This was so despite earlier vacillation, and times of forgetfulness and confusion. The court also stressed the legal presumption of competence, and found that there was insufficient evidence to show that Mrs. Candura's areas of forgetfulness and confusion in any way impaired her ability to understand the consequences of her decision.

The *Candura* court drew on the competency to contract background in a number of ways. It recognized the legal presumption of competency, and noted that the question of legal competency is a question of fact.[10] Most importantly, the court employed the 'understanding test' for determining competency by looking at the patient's capacity to appreciate the nature and consequences of her action. The court took the position that the patient's mental disabilities were not in themselves determinative of her incompetence; rather, to justify such a finding, the disabilities would have to have been found to impair her ability to understand.

In re Quackenbush [16] is another case where a patient was found competent to refuse an amputation of a gangrenous limb despite periods of disorientation. This case came to court when the hospital, alleging the patient to be incompetent, petitioned the court for the appointment of a guardian to consent to the amputation. Although the patient exhibited some fluctuations in mental lucidity, these were found insufficient to affect his competency to make an informed decision. The patient was found to understand the nature and extent of his physical condition, the risks involved with or without the proposed operation, and to appreciate fully the magnitude of his illness. Thus, here too, the court used the 'understanding test' to find the patient competent.

A test closely related to the 'insane delusion test' was employed by a Pennsylvania court in *In re Yetter* [17]. There the court upheld a patient's refusal to submit either to a surgical biopsy or to any additional corrective surgery that might be needed for breast cancer. Mrs. Yetter was, at the time, committed to a mental hospital and some of her reasons for refusal were clearly delusional. Citing an earlier Pennsylvania case [27], the court stressed that mere commitment to a state hospital does not destroy a person's competency. That must be examined on a case-by-case basis. The court employed an understanding test in its finding that Mrs. Yetter was lucid, rational, and appeared to understand that the possible consequences of her refusal included death. In addition, the court addressed the difficulty presented by the fact that Mrs. Yetter's refusal was accompanied by delusions consistent with her mental illness.[11] It upheld her competency, however, by finding that her delusions were not the primary reason that she rejected the treatment. It also noted that a refusal based on fear may be irrational and foolish, but does not warrant a finding of incompetence.

The *Yetter* court expressed an unwillingness to override the patient's decision just because *some* of her reasons were delusional. Such a finding is consistent with the 'insane delusion test' used in the earlier contracts cases. There, a delusion did not affect competency to contract unless the transaction in question was motivated by that delusion. Here, Mrs. Yetter's delusions were not determinative of her competency to refuse treatment where the court found that her decision was not a *product* of those delusions.

Department of Human Services v. Northern [6] is a case where a patient's irrational beliefs *were* found to affect her competency to refuse medical treatment because they affected her ability to understand her condition. The patient was generally lucid and sound of mind, but she was found nonetheless to be incompetent to refuse the amputation of her gangrenous feet. Apparently she did not appreciate the danger she was in and she refused to accept the impossibility of both living and keeping her feet. The court found the patient's denial of this unpleasant reality a 'delusion' which rendered her incapable of making a rational decision with respect to the proposed surgery. Unlike the delusions in the *Yetter* case, this patient's delusions were found to affect her competency to make this particular decision because they were felt to be the motivating force behind the decision.

The *Northern* case has been criticized for adopting "an unduly narrow test of competence" [60]. It should be noted, however, that the decision is basically consistent with the standards established in the competency to contract area. The crucial inquiry involves understanding and appreciating the nature and effect of the *particular* transaction at issue. Because the patient in *Northern* would not accept the seriousness of her condition, she was unable to understand that the consequence of her decision to refuse amputation would be her death. It did not matter that she was found to be lucid and sane in all other respects. She was found to be incompetent to decide this one (albeit large) issue. Similarly, in applying the 'insane delusion test', the patient's delusions were the motivating force behind her refusal.

A problem that often arises in the competency to refuse treatment cases is what has been referred to as the "outcome approach trap" ([38], p. 571). This is when a finding of incompetency is based simply on the fact that the court feels that the patient made an unreasonable decision, rather than on his capacity to appreciate the factors that go into that decision. The focus then switches to the decision itself and away from the person who made it. The use of this type of test for determining competency has been criticized by a number of commentators, some of whom have claimed that it is probably used more often than is admitted by doctors and courts [61]. This problem arose in the *Candura* trial court and it has arisen as well in a number of blood transfusion cases ([1], [32]).

While the 'outcome approach' for determining competency to refuse medical treatment can and should be criticized on a number of grounds, it should at least be noted that such a test is closely related to the 'inarticulate premise' which Milton Green wrote about in the competency to contract area as early as 1944 [47]. Incompetency to contract, according to Green, is often judged not by the actual incapacity of the contracting party, but by the abnormality or unfairness of the transaction in question.

The tests for determining incompetency to contract were developed within a context of concern over distinguishing between those adults who actually had the capacity to make contracts and those who did not. By contrast, at common law *all* children were deemed to be incompetent to contract ([52], pp. 386–394). There have always been certain distinctions made between those contracts made by minors that might be enforced by the child and those that can not, and there have been differences among the states for some time in the relevant age of majority, but traditionally there has not been concern to distinguish among the actual mental capacities of particular children. The *Restatement (Second) of Contracts* conforms to this basic pattern by providing in Section 14 that:

Unless a statute provides otherwise, a natural person has the capacity to incur only voidable contractual duties until the beginning of the day before the person's eighteenth birthday [37].

The question of the competency of minors to make medical treatment decisions has tended, as well, to be decided without concern for their varying actual capacity. Minors are legally incompetent; they are presumed to be incapable of making important decisions on their own [64]. Thus under the general common law rule, minors are deemed to be incapable of making medical treatment decisions [44]. In most states, the age under which incompetency to consent is presumed is 18 or 19 ([50], [44]). Where minors are considered to be incompetent to consent, their parents have legal authority to make decisions for them.[12]

The requirement of parental consent for the medical care of minors has been challenged, however, as overly harsh [63], and many now question the traditional presumption that children are incapable of making such decisions (see [64]). Many argue, for example, that minors today are more sophisticated and mature than they were in the past when the general rule was formed. As a result, both judicial and statutory exceptions to the general rule have arisen.

The most interesting and significant departure from the general rule is the judicially created 'mature minor' doctrine. This doctrine allows some minors to consent to their own non-emergency medical care in at least some cases [63]. The cases generally involve a minor who is 'mature' — near the age of majority and of sufficient mental capacity to understand the nature and importance of the proposed medical procedure — and treatment which is undertaken for his benefit [63]. Most significant is the judicial definition of 'maturity'. It closely approximates the standard used to determine competency in adults and indicates a judicial willingness to consider the actual competency of particular young people.

The Supreme Court recognized the mature minor doctrine in *Bellotti v. Baird* [2], where it upheld the right of a minor to consent to an abortion if she had the capacity to make an informed decision. An earlier lower court case that invoked the doctrine was *Younts v. St. Francis Hospital and School of Nursing, Inc.* [35]. There a 17-year-old was held to be "mature enough to understand the nature and consequences and to knowingly consent to the beneficial surgical procedure. Similarly, in *Smith v. Seibly* [30], in determining whether a married 18-year-old minor was competent to consent to a vasectomy, the court regarded his mental competency as a question of fact and asked whether he fully understood the ramifications, implications, and probable consequences of the procedure.[13]

The judicially created mature minor rule is not the only exception to the general rule that minors are incompetent to consent to medical treatment decisions. There are some statutory exceptions as well. These statutes take one of two general forms. Either they give minors authority to consent to all treatment after a certain age, or they set an age at which minors can consent only to specific forms of medical treatment ([62]), [44]). According to the New York Public Health law, for example, any person 18 or older who is married or is the parent of a child may give effective medical consent for himself [62]. State statutes that authorize minors to consent to specific forms of medical treatment usually involve birth control, pregnancy, venereal disease, or substance abuse [44]. These are areas where it is felt that minors may not secure needed assistance if they are forced to involve their parents. Only the mature minor rule, however, purports actually to examine the competency of the particular minor patient to consent to medical treatment.

It is evident that there are many similarities between the tests used to determine competency to contract and competency to make medical treatment decisions. This has likely occurred in part because the consent required for medical treatment gives the physician-patient relationship a somewhat contractual flavor. Caution is urged, however, against taking this analogy too far. While there are obvious similarities, there are also many differences between the physician-patient relationship and the relationship between parties to an ordinary business transaction [65]. For example, a standard business contract often involves competing parties who appear to have relatively equal bargaining power. A typical physician-patient relationship is quite different. The law may well be on the verge of developing standards of competency to make medical treatment decisions which take account of these differences. It seems clear, at least, that it makes sense for it to begin to move in that direction.

University of Michigan Law School
Ann Arbor, Michigan

Georgetown University Law Center
Washington, D.C.

NOTES

[1] This article was commissioned and originally written in 1985. In updating our research, we find that since then there has been surprisingly little judicial attention paid to establishing standards for determining the competency of patients to make medical treatment decisions ([31], p. 496 n. 26) ([58], [59]). To the extent that the question has been litigated, courts have generally used some version of the understanding test described infra at p. 150 (see, e.g., [12], [21], [54]).

[2] An American Law Reports Annotation entitled, 'Mental Competence of Patient to Consent to Surgical Operation or Medical Treatment', noted in 1967 that very few cases had been found within the subject's scope [39]. By the 1970s, however, a number of cases had arisen concerning competency to refuse medical treatment ([6], [16], [17], [19], [22]).

[3] Alexander and Szasz have attacked this assumption. They argue that the law should

not recognize mental incompetency as a ground for avoiding contracts because to do so impinges on a person's ability to determine the course of his own life [36].

[4] Various courts embellished the basic test with so many qualifying adjectives that one leading commentator referred to it as "ambiguous, self-contradictory and practically meaningless" ([45], p. 147).

[5] The insane delusion test is not limited to contracts. A will may be invalidated if its content were motivated by an insane delusion ([38], p. 580 n. 63). In one case, for example, a will was set aside where the testator disinherited his wife because of his deluded belief that she had been unfaithful to him [15].

[6] Quoting the *Restatement (Second) of Contracts*, section 15, comment b [37].

[7] The competency assessment is not determinative of whether a patient will be treated. There may be other overriding interests that lead to a competent patient's being treated without his consent. For example, a prison inmate may not, in effect, commit suicide by refusing needed dialysis [5]. Alternatively, an incompetent patient's expressed desires may well be fulfilled if these are found to be in his "best interests". (See [60].)

[8] By contrast, Annas and Densberger refer to the standard as "probably the most precise concept of competence that we will be able to develop" ([38], p. 572).

[9] Given the likely genesis of physicians' recent sensitivity to the matter of patient competence, it seems surprising that there are not many examples of doctors acting to protect themselves by seeking the appointment of a guardian to authorize the treatment of an agreeable but possibly incompetent patient. (See [60], pp. 139—143).

[10] "The principal question is whether the facts established by the evidence justify a conclusion of legal incompetence" ([19], p. 1235).

[11] Mrs. Yetter, who was 60-years-old, believed that the operation would affect her ability to have babies and would preclude the movie career she dreamed of having [17].

[12] While these are the *legal* standards, the following quotation from an American Psychological Association position paper gives some evidence of how they are actually followed in practice:

> Despite parents' legal authority to make most decisions for their children, the preferences and concerns of children regarding involvement in treatment are sought and considered. In cases where minors are authorized by law to provide independent consent for a particular treatment, their desire to decide independently is honored. However, where appropriate, such minors are encouraged to involve their parents in treatment decisions [67].

[13] More recently an Illinois court held that a mature minor has a common law right to consent to or refuse medical treatment, even if the refusal would result in death ([13], [20]).

BIBLIOGRAPHY

Cases
1. Application of President and Directors of Georgetown College, 331 F.2d 1000 (D.C. Cir. 1964).
2. Bellotti v. Baird, 443 U.S. 622 (1979).

3. Border v. Kelso, 293 Ill. 175, 127 N.E. 337 (1920).
4. Cobbs v. Grant, 8 Cal. 3d 229, 502 P.2d 1 (1972).
5. Commission of Corrections v. Myers, 399 N.E. 2d 452 (Mass. 1979).
6. Department of Human Services v. Northern, 563 S.W. 2d 197 (Tenn. Ct. of Appeals 1978).
7. Dexter v. Hall, 82 U.S. (15 Wall.) 9, 20 (1872).
8. Dunham v. Wright, 423 F.2d 940 (3d Cir. 1970).
9. Farber v. Sweet Style Manufacturing Co., 40 Misc. 2d 212, 242 N.Y.S.2d 763 (N.Y. Sup. Ct. 1963).
10. Grannum v. Berard, 70 Wash. 2d 304, 422 P.2d 812 (1967).
11. Howe v. Massachusetts, 99 Mass. 88, 98—99 (1868).
12. In re A.C., 573 A.2d 1235 (D.C. App. 1990).
13. In re E.G., a minor, 133 Ill.2d 08, 549 N.E. 2d 322 (1989).
14. In re Holloway's Estate, 195 Cal. 711, 733, 235 P. 1012, 1021 (1925).
15. In re Honigman's Will, 8 N.Y.2d 244, 168 N.E.2d 676, 203 N.Y.S.2d 859 (1960).
16. In re Quackenbush, 156 N.J. Super. 282, 383 A.2d 785 (1978).
17. In re Yetter, 62 Pa. D & C 2d 619 (1973).
18. Krasner v. Berk, 366 Mass. 464, 319 N.E.2d 897 (1974).
19. Lane v. Candura, 376 N.E.2d 1232 (Mass. App. 1978).
20. Long Island Jewish M. Ctr., 557 N.Y.S.2d 239 (Sup. 1990).
21. Matter of Farrell, 529 A.2d 404 (N.J. 1987).
22. Matter of Schiller, 148 N.J. Super. 168, 372 A.2d 360 (1977).
23. Mohr v. Williams, 95 Minn. 261, 104 N.W. 12 (1905).
24. Moore v. Webb, 345 S.W.2d 239 (Mo. 1961).
25. Mutual Life Ins. Co. v. Hunt, 79 N.Y. 541, 545 (1880).
26. Ortelere v. Teachers' Retirement Board, 25 N.Y.2d 196, 303 N.Y.S.2d 362, 250 N.E.2d 460 (1969).
27. Ryman's Case, 139 Pa. Superior Ct. 212 (1939).
28. Schloendorff v. Society Hospital, 211 N.Y. 125, 105 N.E. 92 (1914).
29. Smith v. Seibly, 72 Wash. 2d 16, 431 P.2d 719 (1967).
30. Thompson v. Leach, 3 Mod. Rep. 301, 87 Eng. Rep. 199 (1690).
31. U.S. v. Charters, 829 F.2d 479 (4th Cir., 1987). (Later overturned en banc at 863 F.2d 302 (1988)).
32. United States v. George, 239 F. Supp. 752 (D. Conn. 1965).
33. Uribe v. Olson, 42 Or. App. 647, 601 F.2d 818 (1979).
34. Willett v. Willett, 333 Mass. 323, 324, 130 N.E.2d 582 (1955).
35. Younts v. St. Francis Hosp. & School of Nursing, Inc., 205 Kan. 292, 469 P.2d 330 (1970).

Articles and Books
36. Alexander, G. and Szasz, T.: 1973, 'From Contract to Status via Psychiatry', *Santa Clara Lawyer* **13**, 537—559.
37. American Law Institute: 1981, *Restatement of the Law Second, Contracts 2d*, A.L.I. Publishers, St. Paul.
38. Annas, G. and Densberger, J.: 1984, 'Competency to Refuse Medical Treatment: Autonomy v. Paternalism', *Toledo Law Review* **15**, 561—596.

39. (Annotation): 1969, 'Mental Competency of Patient to Consent to Surgical Operation or Medical Treatment', *American Law Reports 3d* **25**, 1439—1443.
40. Calamari, J. and Perillo, J.: 1977, *The Law of Contracts*, 2nd ed., West Publishing Co., St. Paul.
41. Capron, A.: 1974, 'Informed Consent in Catastrophic Disease Research and Treatment', *University of Pennsylvania Law Review* **123**, 340—438.
42. Cook, W.: 1921, 'Mental Deficiency and the Law of Contract', *Columbia Law Review* **21**, 424—441.
43. Dawson, J., Harvey, W. and Henderson, S.: 1982, *Cases and Comment on Contracts*, 4th ed., Foundation Press, New York.
44. Dresser, R.: 1984, 'Feeding the Hunger Artist: Legal Issues in Treating Anorexia Nervosa', *Wisconsin Law Review*, 297—374.
45. Green, M.: 1941, 'Judicial Tests of Mental Incompetency', *Missouri Law Review* **6**, 141—165.
46. Green, M.: 1943, 'The Operative Effect of Mental Incompetency on Agreements and Wills', *Texas Law Review* **21**, 554—589.
47. Green, M.: 1944, 'Proof of Mental Incompetency and the Unexpressed Major Premise', *Yale Law Journal* **53**, 271—311.
48. Grotius, H.: 1646, *Of the Rights of War and Peace*, Book II.
49. Keeton, W. *et al.*: 1984, *Prosser and Keeton on the Law of Torts*, West Publishing Co., St. Paul.
50. Keith-Speigel, P.: 1981, 'Children and Consent to Participate in Research', in *Children's Competency to Consent*, Melton (*et al.*), pp. 179—207.
51. Kelly, W.: 1960, 'The Physician, the Patient, and the Consent', *Kansas Law Review* **8**, 405—434.
52. Leake, S. M.: 1912, *The Law of Contracts* (6th ed.), Stevens and Sons, Lmtd., London.
53. McCoid, A.: 1957, 'A Reappraisal of Liability for Unauthorized Medical Treatment', *Minnesota Law Review* **41**, 381—434.
54. McCrary, S. V. and Walman, A. T.: 1990, 'Procedural Paternalism in Competency Determinations', *Law, Medicine & Heath Care* **18**, 108—113.
55. Meisel, A.: 1979, 'The "Exceptions" to the Informed Consent Doctrine: Striking a Balance between Competing Values in Medical Decisionmaking', *Wisconsin Law Review*, 413—488.
56. (Note): 1974, 'Informed Consent and the Dying Patient', *Yale Law Journal* **83**, 1632—1664.
57. (Note): 1959, 'Mental Illness and Contracts', *Michigan Law Review* **57**, 1020—1118.
58. (Note): 1990, 'Determining Patient Competency in Treatment Refusal Cases', *Georgia Law Review* **24**, 733—757.
59. Perlin, M. L.: 1990, 'Are Courts Competent to Decide Competency Questions? Stripping the Facade from United States v. Charters', *Kansas Law Review* **38**, 957—1001.
60. Robertson, J.: 1991, 'The Geography of Competency', in this volume, pp. 127—148.
61. Roth, L., Meisel, A. and Lidz, C.: 1977, 'Tests of Competency to Consent to Treatment', *American Journal of Psychiatry* **134**, 279—284.

62. Wadlington, W., Whitebread, C. and Davis, S.: 1983, *Cases on Children in the Legal System*, Foundation Press, New York.
63. Wadlington, W.: 1973, 'Minors and Health Care: The Age of Consent', *Osgoode Hall Law Journal* **11**, 115–125.
64. Weithorn, L.: 1983, 'Children's Capacity to Decide About Participation in Research', *IRB* 5.
65. White, P.: 1983, 'The Common Law as a Model of the Patient-Physician Relationship: A Response to Professor Brody', in E. E. Shelp (ed.), *The Clinical Encounter*, D. Reidel Publishing Com., Dordrecht, Holland, pp. 133–139.
66. Zimring, F.: 1982, *The Changing Legal World of Adolescence*, Free Press, New York, pp. 91–93.

Unpublished Paper
67. 'Position Statement: Standards Regarding Consent for Treatment and Research Involving Children', *American Psychological Association* Third Draft: January, 1984.

PART IV

CASE STUDIES: SEEKING INSIGHT FROM APPLICATION

EUGENE V. BOISAUBIN

COMPETENCY JUDGMENTS: CASE STUDIES FROM THE INTERNIST'S PERSPECTIVE

Occasional cases in the clinical setting involve ethical conflicts requiring the expertise of ethicists, attorneys, and psychiatrists. Physicians, properly trained in the basic techniques of patient interviewing and physical and mental examination, can, however, recognize dementia and organic disease states affecting patients' competence to decide in matters regarding their treatment. This may require a greater emphasis in medical education on those states reflecting underlying mental diseases such as depression and schizophrenia. Given proper instructions, however, physicians may be trained to make proper psychiatric diagnoses. When there are questions, psychiatric consults may be solicited. Only in matters of irresolvable conflict among health care professionals, patients, and interested parties should legal advice be sought.

This essay presents four cases that can be analyzed and resolved by a well-trained primary care clinician. The cases focus generally on patients' competency to consent to treatment and internists' judgments regarding this ability. Here competence is understood as a capacity to act voluntarily with sufficient information. The most meaningful way to test for patient competence in the clinical setting is to determine whether the patient actually understands the information necessary to provide informed consent. If a positive determination can be made, a patient may be said to have the capacity or competence to decide upon a future course of treatment.

The first case discusses a patient with a serious disease for which treatment, or at least careful observation, may prevent a fatal outcome. The patient appears initially to be competent, but fear and denial are seen by the clinician to distort the patient's decision-making capacities. In the second case, a fatal disease is present that may only be palliated. Fear and physiologic incompetence are strongly suspected and honoring the patient's request to forego treatment may result in premature death. In the third case, a patient has a serious disease for which treatment and cure are available and competency may be altered by both physiologic and psychiatric factors. Finally, in the fourth case, a competent and uncooperative patient has a potentially serious disease that is controllable only by long-term therapy.

CASE I

A 45-year-old businessman who has been complaining of "gas pains" for 6 hours is brought reluctantly to the Emergency Center by his wife. His subsequent cardiogram shows a possibility of an early myocardial infarction (heart attack) and the physician recommends hospitalization for monitoring. The patient refuses, saying he needs to return to his office to complete a business transaction.

Discussion

At first blush, this patient might appear to be competent to decide to refuse medical care. Several elements of his history, however, should be disquieting to the attentive physician. First is the chief complaint or perceived symptom of the patient. Although an acute myocardial infarction in the diaphragmatic region may present as gas or nausea, the patient may choose to deny other associated symptoms, such as squeezing chest pain, malaise, or shortness of breath, thereby postponing medical attention. A great danger to the patient with myocardial infarction is the tendency to deny or procrastinate with the early symptoms at a time when risk of death is greatest. This is the case even though, today, most reasonably educated middle-aged men think of a heart problem when chest pain occurs.

The physician needs to make a clinical decision and recommendation for care in this case. How strongly he or she believes in and pursues that decision medically depends upon a number of variables, including the weight of the historical information, physical examination and ancillary data. If the physician is reasonably convinced of the correctness of the diagnosis, the accepted standard of medical care in America today would be hospital admission with monitoring, preferably in an intensive care setting. In this case, let us assume that the cardiogram is very suggestive of a myocardial infarction, the patient's history is compatible, and immediate hospitalization with monitoring is the best correct medical decision.

Let us further assume that, despite this professional recommendation, the patient decides to forego care and to return to work. The physician has several options. At one end of the spectrum, the physician might try to force the patient to stay by challenging the patient's decision-making ability on the basis of 'physiologic incompetence' — that is, by arguing that the patient has a metabolic problem secondary to his infarction, thus compromising his decision-making ability. If the patient is hypoxic, hypotensive, or acidotic, this argument could be made with some justification. This physiologic argument would be strengthened if, in addition, the physician found other evidence of general cerebral dysfunction such as confusion, disorientation, or memory loss. However, physiologic incompetence is usually not present with myocardial infarction and other reasons for the patient's decision would have to be addressed. The second, more probable basis for an alteration in the patient's decision-making ability may be a less obvious impairment — emotions such as fear or anger, or a more complex process such as denial. Suggesting this are elements already present, such as a hesitancy to submit to initial assessment, possible misinterpretation of symptoms, and the assigning priority of work-place duties over possible death. This issue should be approached by further questions concerning the basis for these fears or apprehensions, followed by explanation and reassurance.

Another option is that the patient has made a truly informed, reflective decision to forego medical care and accepts the consequences. If the physician judges that these criteria of rational understanding and appreciation of the consequences have been met, that no major factors of organicity or emotional impairment were present, and that all avenues of reasonable persuasion, including the involvement of family members have been tried, the physician will be obligated to respect the patient's wishes.

There is yet one other option. If the physician judges that issues such as denial or fear are present, and if the patient is truly ambivalent about his decision to leave, a measure of strong altruistic manipulation may be warranted. A physician may proceed to tell the patient calmly but directly that he must be admitted and then proceed with the appropriate clinical action. Of course, the patient has the right, and in this case the ability, to walk away at any time. Given the alternative actions the patient may take, his expressed wishes are tested. If the patient does decide to leave, he should be counseled and encouraged to return at any time without sanction.

CASE II

A 40-year-old man has been diagnosed two months previously as having inoperable lung cancer. He is currently receiving radiotherapy which is palliative. Over the past two weeks he has been getting progressively short of breath, and an endotracheal tube is placed through his mouth to his trachea for breathing support. He is admitted to the Medical Intensive Care Unit and, upon arrival, gestures that he wants the tube removed. This gesture is also clarified by a written request on a note pad.

Discussion

Let us assume that the tube was initially placed because the patient was hypoxic with great difficulty breathing and, without assisted ventilation would have soon died. Assume we do not know if the patient had expressed any initial objection to the intubation, had expressed a previous desire for a particular course of clinical action to his primary physician or family, or even had written an advance directive (i.e., a living will). If either of the last two circumstances were known to be present, this information will (and should) have a major impact upon the subsequent clinical action. However, an objection to the initial intubation probably would have been overridden by most physicians because of the assumed temporary incompetence of the patient secondary to hypoxia and the emergent nature of the situation. At this point, the physician's dilemma is whether to honor the patient's wishes. Most physicians, in my experience, would not honor his wishes at this time for the reasons present when initial intubation was performed. These reasons are founded upon evidence in favor of, for example, incompetence secondary to brain hypoxia or another yet undiscovered metabolic problem; fear, anxiety, depression, or merely inadequate knowledge about this condition, its outcome and his options. After insertion of the tube, no immediate additional harm is probable, although removal would possibly result in substantial harm, i.e., death. The prudent medical action would be to allow the tube to be left in place until greater assurance about the competence of the patient can be established. Medically, a period of perhaps 24 hours would allow his metabolic status to correct or stablize and contributing issues such as fear and anxiety might subside or be directly addressed. At 24 hours, then, two options would be present. First, if the patient's hypoxia had improved, removal of the endotracheal tube would be medically indicated and the conflict at hand would disappear. But if brief trials without respiratory support suggest that permanent removal would result in respiratory failure and death, the conflict of patient choice versus professional treatment reappears. Assuming that there are no persisting metabolic problems, the next issue is whether the patient understands the nature and implications of his decision. Does he understand the nature of his condition, its ultimate outcome, and his chances for a reasonable interval of existence? Would his decision to stop ventilator support certainly, probably, or only possibly result in death? Indeed, this is a very difficult medical prognostication; nevertheless, the physician must give the patient a prognosis based on present clinical knowledge. Following the communication of this prognosis, the physician might wish to encourage the patient to speak with his family members, friends, or clergy for advice. Regardless of what transpires, all concerned will probably have doubts about the setting for the decision. An ICU by its very design and use is a poor location for reflective decision-making. Its glaring, noisy, and frenetic pace creates an unnatural, unrelenting stimulus for patients and staff. Normal day-night cycles are nonexistent and sleep, when it comes, results from total exhaustion. Patients often describe a sense of powerlessness as they are surrounded by and dependent upon machinery that monitors countless functions and may even sustain their very existence. One of the life-sustaining interventions, intubation, is particularly intrusive, frightening, and continually uncomfortable, if not painful. Patient interaction with the ICU professional staff is usually curt, task related, and void of substantial emotional content. Contact with family members is restricted, usually to ten minutes which is hardly conducive to relaxed, intimate communication.

Despite these important issues, a decision must be made. I contend that at 24 hours, if the patient is metabolically stable, competent by mental status exam, has had time for reflection with family and friends, and understands the implications of his decision and his option to change his mind, the endotracheal tube may be removed with moral impunity.

CASE III

A 50-year-old woman with a reported history of paranoid schizophrenia presents to a clinic with a chief complaint of "nerves" of several months' duration. On physical examination, she is found to be tachycardic, anxious, and with an enlarged thyroid gland. Her subsequent blood tests confirm the clinical diagnosis of hyperthyroidism. Her condition is explained to her and therapy with intravenous radioactive iodine is recommended. She refuses to allow therapy, stating that the radiation will certainly "poison" her body and possibly "even kill her".

Discussion

In this case, as in the other two, we have a direct conflict between the physician's recommendations for treatment and the patient's expressed wishes. I begin with the premise that any competent patient has the right to refuse treatment. If that right is to be overridden, the physician must be able to identify factors affecting competence that would justify such an action. This case presents several such factors. First is the patient's previous diagnosis of paranoid schizophrenia. Although the historical basis of any diagnosis can and should often be questioned, the more important question is whether that condition currently exists and affects clinical decision-making. A series of questions from the physician to the patient concerning disturbances and experiences such as auditory hallucinations, bodily delusions, and observations of abnormal modes of expression and language will clarify the existence and extent of schizophrenia. Obviously, a psychiatrist is the best consultant to assist with this analysis.

But, even if some or all of these questions resulted in answers compatible with schizophrenia, the issue is still not resolved. The physician must continue to focus specifically upon whether the patient's perception of therapy as poison represents a schizophrenic delusion, a misinterpretation of medical reality perhaps intensified by fear, or an informed rational decision based upon the patient's unique set of values. This analysis is most easily resolved in a clinical setting by intensely questioning the patient's rationale or motives for the decision, i.e., why do you believe the recommended treatment will poison you? If the patient's answer suggests a delusional perception that seems to be part of a larger delusional state, incompetence by virtue of schizophrenia is probable. Careful probing of the patient's rationale will also allow identification of the second possible factor, that is, misinterpretation. Does the patient fully understand the nature of her disorder, its manifestations, natural history, and outcome? Has the nature of the recommended radioactive therapy been explained, including its possible side effects and long-term safety? Ignorance and fear of this particular, or any other, treatment modality can often be overcome with education and reassurance. Last, is the patient's decision based not upon schizophrenia, ignorance or fear, but rather upon a logical set of personal values? Ever since Three Mile Island, many rational Americans have had grave concerns about the unknown risks of radiation in any form. No honest physician can fully guarantee the ultimate safety of internal radiation therapy and some patients may view physicians as members of a larger conspiracy in the scientific community to foist hazardous treatment upon the unsuspecting public. Although, for example, most scientists and physicians would view fear of drinking water fluoridation as 'irrational', patients like all members of our society have the option and right steadfastly to maintain views that the scientific community may find naive or incorrect.

Schizophrenia aside, a second factor may also be present to alter competence. That factor, as in the other cases, is an abnormality of cerebral function secondary to an underlying disease. In this case, the hyperactive thyroid gland is producing thyroxin and other hormonal metabolites that in high concentrations can alter normal thought processes and even create a delusional state similar to schizophrenia. Patients with underlying mental disease may even be more susceptible to these metabolic effects than otherwise normal patients. A high output hormonal state is very probable since the patient is already tachycardic and anxious (two premier signs of marked thyroid hormone excess). With the possibilities present of metabolic effects, the clinician would want to ask questions to determine if there has been an impairment of judgment. These questions would address orientation to person, place and time, recent and remote memory, attention and concentration. Even if the patient showed no evidence of global dysfunction, specific questions again about the patient's understanding of the nature of her condition, proposed treatment, and options must be addressed. Inability to answer these questions in a normal fashion would support the diagnosis of thyroid-related encephalopathy. Hyperthyroidism *per se* would be insufficient to declare incompetence but a thyrotoxic state, coupled with clear findings of encephalopathy, would be sufficient cause.

At this point, however, some responsibility for resolution of the problem should be shifted to the physician. Although he or she had recommended radioactive iodine as the 'best possible treatment', two other modalities, long-term drug therapy and surgery, are reasonable options. Although the medical pros and cons of each modality are complex and beyond the scope of this discussion, the point is that the physician, as 'gatekeeper' of therapeutic options, must remain open-minded and flexible to alternative plans. The so-called 'best' decisions for most patients may have to yield to a second best choice for an individual patient. A patient might readily accept the risk of surgery over the risks of radiation exposure. Last is the possibility that competence could be restored before thyroid therapy was initiated. If the patient were found to be actively schizophrenic and competency to decide regarding the thyroid therapy was impaired or even uncertain, drug therapy for the schizophrenia could be instituted first. If the patient responded to psychotropic medication, the chances for improved decision-making concerning thyroid therapy should also improve.

In summary, this patient presents a complex case of possibly interacting factors affecting competence. If after careful assessment of mental status I felt this patient were actively schizophrenic, I would institute anti-psychotic drug therapy and temporize the thyroid condition with oral drugs. If the patient responded with an improved mental status and competency were restored, I would negotiate all three thyroid therapeutic options with the patient. If incompetence were secondary to metabolic encephalopathy, I would also try to treat initially with oral anti-thyroid drugs until competency was returned and negotiate regarding further treatment. Last, if competency were present, I would reeducate the patient about all three treatment modalities, express my professional preference, and ask her to make a decision about which, if any, of the options she desired.

CASE IV

A 45-year-old asymptomatic housewife with a college education presents to a physician's office after being referred for follow-up of elevated blood pressure detected at a shopping center. After two office visits, her blood pressure is still in the range of 180/105. A history, physical exam, and workup fails to reveal any correctable cause for her problem and the physician recommends moderate reduction in salt ingestion and use of an antihypertensive drug as a means of normalizing her blood pressure. The patient politely refuses, however, saying that she believes in a more "natural" approach to therapy and would rather follow a diet high in certain herbs and grains recommended in a local health food store.

Discussion

This patient presents a dilemma found in outpatient medicine. She expresses a desire to follow a mode of therapy that is viewed as unconventional and ineffective by the medical community. Although some physicians might wish to view such a patient as incompetent, the odds for this woman having total control of her mental faculties are good. First, the fact that she has voluntarily followed up her medical problem and participated in a workup to this point suggests competence. We will also assume that during the course of her workup, no indication was found of obvious metabolic or psychiatric disease. As a last check for delusional ideas, but primarily to clarify the rationale behind her decision, the physician must ask additional questions.

Does she fully understand her disease and the potential for harm if left unattended? Does she know that her disease may remain asymptomatic until irreparable harm has occurred? Does she realize that drug therapy must be taken indefinitely and may have side effects, but is efficacious? Is she willing to consider another drug rather than the one prescribed? This patient has not only the right to know about her disease, but the education and demonstrated interest to understand the advantages and disadvantages of medical therapy presented in a clear, unbiased manner.

Although most discussions of patient compliance focus unilaterally on the virtue or sins of the patient in adhering to a treatment regimen, Al Jonsen [1] has argued that the issue is clearly bilateral. How successful has the physician been in understanding the patient's motives, comprehension and fears? Has a technique of education been used that is compatible with the patient's education, life setting, and needs? Or has the physician already committed a sin of coercion by overstating the risks of hypertension ("you will surely have a stroke") or understating the risks of treatment ("you will have no problem at all with this drug")? The ultimate purpose of patient education is not to make the patient do what the doctor believes, but give patients insight into their conditions and freedom to act as they desire.

If after extensive discussion the patient adamantly refuses the prescribed regimen, the physician has several options. The first might be to terminate this physician/patient relationship and refer the patient to another health care provider. If the physician truly believed that this prescribed drug regimen was the only acceptable, professionally conscionable treatment modality, such an action may be warranted. The physician might argue that successful treatment of hypertension requires belief by the patient in the proposed regimen as a requisite for prolonged daily compliance, and without this adherence, a continued professional relationship would be futile. Conditions like hypertension, because of their asymptomatic and protracted nature, depend almost totally on patient compliance for successful treatment. If the doctor wanted to terminate the relationship, however, referral to another physician would be difficult, since most practitioners would recommend a similar drug protocol. Several avenues of therapeutic compromise, however, are possible. Assuming the herbal therapy is harmless, the physician might be able to negotiate a plan with the patient that includes both medically conventional and herbal therapy. As an extension, the physician might even contact the health store proprietor in an attempt to enlist support for the plan. Many physicians on principle would initially refuse to follow this plan, but might be convinced of its virtue by a consequentialist argument that emphasizes the ultimately harmless and beneficial outcome. Assuming that the hypertension was not immediately threatening to the patient's health, an even more therapeutically liberal physician might recommend that the patient first try the herbal therapy and monitor the results. This approach has the advantage of maintaining the patient's autonomy and allowing her to follow and decide about the results, as she chooses.

If the patient still declines to follow all proposed alternatives of care and the physician has carefully applied his or her best educational and self-analytical skills to the situation, the endeavor can be abandoned. As always, the physician should leave the door open for the patient to return at any time for treatment or consultation. What health professionals call non-compliance may simply reflect a different but more important set of life values for the patient.

Baylor College of Medicine
Houston, Texas

BIBLIOGRAPHY

1. Jonsen, Albert R.: 1979, 'Ethical Issues in Compliance', in R. B. Haynes *et al.* (eds.), *Compliance in Health Care*, Johns Hopkins University Press, Baltimore, Maryland, pp. 113—120.

MARK PERL

COMPETENCY JUDGMENTS: CASE STUDIES FROM THE PSYCHIATRIST'S PERSPECTIVE

I. INTRODUCTION

Psychiatrists at times are consulted by colleagues in other specialties to assist in determining a patient's capacity to give (or withhold) consent for medical treatment. Dr. Boisaubin rightly points out that in most instances, a properly trained physician alone can determine whether a patient is making informed choices about medical care. Using a combination of common sense, basic interviewing skills, and fundamental psychiatric concepts, practitioners usually know when patients have grasped the medical facts presented to them, and when they are making a choice based on that knowledge.

When a psychiatrist is called in, the case may be fairly clear-cut, and the psychiatrist is merely asked to confirm the primary physician's impressions. On other occasions, there may be genuine doubt as to whether the patient is acting as an informed, autonomous person. The psychiatrist is typically asked to evaluate the patient's mental state, and determine 'competency' to give informed consent.

It is important to remind ourselves that 'competency' (in the narrow, technical sense) is a legal rather than a medical/psychiatric concept. Even though most cases present no problem, and physicians alone decide about competency, dubious cases are referred to the courts as the final authority. Criteria for competency vary between different authorities and jurisdictions, and between different treatment situations. There are no unanimous standards for competency, although various tests and definitions have been devised ([1], [3], [4], [6]). Furthermore, as Appelbaum and Roth point out [1], competency is not "... a fixed attribute of an individual, a characteristic in itself with an inherent stability", but is in fact "... a set of deductions from a variety of clinical data that can be as subject to influence and change as the more basic mental attributes on which it is based". These authors see the psychiatrist as providing "... a factual basis for the legal determination that corresponds to the patient's clinical state".

It is instructive to review these tests of competency as represented in

case law and the legal and psychiatric literature ([1], [2], [3], [4], [6]). Such tests are in a state of evolution; the criteria overlap to some degree, but at times appear to compete with each other. Roth *et al.* [6] enumerate five categories, each of which has at times been considered a sufficient condition for competency to make medical decisions:
— evidencing a choice (i.e., the patient's choice is acceptable so long as he or she merely makes it clear what the decision is)
— 'reasonableness' of choice (i.e., making a choice whose outcome would be acceptable to a 'reasonable' person)
— 'rational' reasons (i.e., whether the patient used 'rational' thought processes to arrive at a decision)
— presence of the ability to understand (i.e., whether the patient could understand the nature of the illness, the benefits and risks of the proposed treatment, and the alternatives)
— presence of actual understanding (i.e., did the patient actually understand the explanations offered)

Yet another criterion is listed by Appelbaum and Roth [2], namely, "appreciation of the nature of the situation". Although this occurs within a discussion of the ability of subjects to consent to research, it is also relevant to the clinical context.

In view of the unclarity in this field, and the lack of unanimity about what constitutes 'competency', it is worth making a distinction between the legal concept (which is not the primary focus of this discussion) and the intuitive, clinical sense we have about patients' abilities to make choices (i.e., competency in the broad sense, for which terms like 'capacity, ability' are useful synonyms).

It is often suggested by clinicians (and implied by Dr. Boisaubin) that fear, sadness, or anger may compromise a patient's competency. Such strong emotions may influence a patient in accepting or rejecting treatment, and in extreme cases may cause him or her to make a 'wrong' or 'irrational' choice, one that would be later regretted. But such emotions are ubiquitous in the medical context, and in human affairs in general. Many decisions and opinions would have to be discounted if those accompanied by strong emotions had to be overruled.

There is no reason to believe that the presence of such feelings in and of itself rules out making informed choices, so long as the patient's cognitive abilities are intact, there are no defects of perception, and he or she is in full possession of the requisite medical facts. (In my experience, this concern — about fear, anxiety, sadness, etc. — is almost never used to question patients' consent to treatment, but only their refusal of treatment.) Invoking strong emotions as a reason to invalidate a patient's decision about treatment would appear to be a subtle form of an *ad hominem* argument. What the patient says and wishes is deemed invalid despite his or her apparently rational reasoning. The clinician focuses not on what the patient is saying, but on antecedent or present emotional states and motives. It is an impossible task (even for a psychiatrist!) to quantify the emotion present, or to judge what degree of 'distortion' may be occurring as a result of the strong feelings experienced.

The argument that patients' wishes or 'unwise' decisions in these situations should not be respected, since they may later be changed, appears unconvincing. Again, this criterion, uncritically applied, would leave very few opinions or decisions unscathed, since changing one's mind (as well as being free to experience one's emotions) is a fundamental requirement of autonomous functioning.

A brief comment about 'denial' is in order. In a recent paper [8], Dr. Earl Shelp and I pointed out that 'denial' is a confusing and imprecise term. It is used to represent two quite different phenomena:

a) The act of negating, disagreeing with what someone else is saying,
b) A hypothetical unconscious defense mechanism, which causes persons to be unaware of certain unpleasant facts.

We pointed out that often physicians use the term in the second sense, when all that has really occurred is negation, that is, 'denial' in the first sense. As we said:

... Granted, the patient is denying in the first sense some evaluation, explanation, description, or recommendation ... (by the physician) ... but he or she is denying in the second sense, that is, using a primitive unconscious defense to avoid reality, only if reality is defined as "what the physician believes to be the fact ...". It certainly would be unwarranted to assume that the patient who disagrees with his or her physician is always out of touch with reality ([8], p. 698).

Because of the confusion of meanings, as well as the difficulty of reliably ascertaining the nature of any purported unconscious process, we proposed abandoning the use of the term 'denial' altogether, and substituting 'minimizing illness'. Certainly patients who minimize illness are not *ipso facto* incapacitated. Persons have a right to minimize illness, even if this results in poor choices.

In what follows I shall address the cases discussed by Dr. Boisaubin in this volume.

CASE I

There is nothing in this brief case presentation that would lead one to believe that the patient is not competent. The patient is undoubtedly minimizing ('denying') both the severity of his symptoms, and the implications for his health. Indeed, from a physician's standpoint, the patient is making an unwise choice. But as we have discussed, this alone does not imply that the patient is incompetent.

Dr. Boisaubin outlines several possible options for the physician in dealing with this situation. Claiming 'physiologic incompetence' — based on some hypothesized brain dysfunction due to hypoxia — would not be appropriate, unless there were definite physiologic abnormalities, accompanied by wide-ranging mental status changes. The patient might appear delirious, with disorientation, memory and concentration impairment, confusion, and possible hallucinations. Such states are possible but not usual in myocardial infarction.

Absent cognitive or perceptual changes, the presence of strong emotions (fear or anger) accompanying the minimization of illness would not seem to constitute a sufficient basis for overriding the patient's decision. Thus, though the patient may indeed be influenced by emotional factors, their presence is not necessarily relevant to the issue of competence. There may be room for discussion with the patient about his feelings, and even efforts at persuasion. It must remain clear at all times, however, that the patient remains free to leave the hospital against medical advice whenever he wishes. A psychiatric consultation may at times be offered to the patient, although such consultation may be viewed as coercive and intrusive by many persons. The purpose of the consultation would be to give the patient the opportunity to explore his feelings about his illness, should he wish to do so.

CASE II

This case highlights the issue of 'physiological incompetence' as a cause of the loss of the capacity to make informed choices. Certainly, in the situation described, hypoxia and other significant metabolic derangements are common. A patient suffering from delirium as a result of such metabolic factors would be confused, disoriented, and would suffer possible perceptual distortions. Physicians would have no hesitation in continuing treatment with a demonstrably delirious patient. If, on the other hand, the patient evidenced full understanding of the medical issues, and no cognitive or perceptual distortions, despite his physiologic problems, then, as Dr. Boisaubin suggests, his decision to cease treatment should be respected.

The only difficulties apparent in this case are the technical difficulties of determining a patient's mental state when communication is a problem. The patient is intubated, and assessing mental state via the written word is arduous. The patient may tire easily, may be unable to write for some reason, or may simply not wish to co-operate. Certainly, the decision to withhold or cease potentially life-saving treatment would not be made lightly. If the patient's mental state could not be assessed with any certainty, one would err on the side of continuing treatment, just as one would initiate life-saving treatment in any emergency situation, where the patient's competency was in some doubt.

One again, fear, sadness, anger, or minimization of illness would not seem in and of themselves to imply incompetence, provided the patient's reality testing was intact. Likewise, the intensive care setting, though distracting and noisy, does not seem to preclude informed decision-making in and of itself.

CASE III

The patient in this case may simply lack sufficient information about radioactive iodine treatment, or she may be delusional. If the former, then she presents no issue of competency; her physicians would simply need to spend time with her carefully explaining the various treatment options before eliciting her decision.

If, on the other hand, the patient is globally delusional and does not understand the medical explanations offered to her, then she would be found to lack the capacity to give informed consent. (It is interesting to note, however, that in many jurisdictions, psychiatric patients are presumed legally competent to make decisions about medical care, even though they may be actively psychotic, and may even be committed to a psychiatric hospital. In some cases they may refuse psychotropic medications while so committed.) The delusions may originate from schizophrenia, from the cerebral effects of thyrotoxicosis, or from a combination of these two processes: the findings with regard to competency would be the same, no matter what the origin of her psychosis.

Assuming the patient were indeed found to be incompetent (and the appropriate legal steps taken to enable treatment to ensue), it would be prudent to follow the course of treatment least objectionable to the patient. Surgeons will only reluctantly operate on unwilling patients, especially in a relatively complicated procedure like thyroid removal. As Dr. Boisaubin points out, compromise arrangements may have to be tried, so as to minimize patient distress while still providing a good (though not perhaps optimal) treatment option. Certainly the least invasive means should be tried first, including measures that would mitigate the patient's psychotic thinking.

CASE IV

There is no question of incompetency in this patient, at least from the case-material presented. The patient's beliefs and values are quite different from those of the doctor, and she may be making unwise decisions, when these are viewed from a medical perspective. Dr. Boisaubin's suggestions about negotiations with the patient in an effort to achieve a compromise arrangement appear to promote caring and mutual respect between patient and physician. Ultimately, the choice about the best course of treatment is the patient's.

Stanford Medical Center
Stanford, California

BIBLIOGRAPHY

1. Appelbaum, P. S. and Roth, L. R.: 1981, 'Clinical Issues in the Assessment of Competency', *American Journal of Psychiatry* **138**, 1462—1467.
2. Appelbaum, P. S. and Roth L. R.: 1982, 'Competency to Consent to Research', *Archives of General Psychiatry* **39**, 951—958.
3. Hoffman, B. F.: 1980, 'Assessing Competence to Consent to Treatment', *Canadian Journal of Psychiatry* **25**, 354—355.
4. Meisel, A., Roth, L. R. and Lidz, C. W.: 1977, 'Toward a Model of the Legal Doctrine of Informed Consent', *American Journal of Psychiatry* **134**, 285—288.
5. Perl, M. and Shelp, E. E.: 1982, 'Psychiatric Consultation Masking Moral Dilemmas in Medicine', *New England Journal of Medicine* **307**, 618—621.
6. Roth, L. R., Meisel, A. and Lidz, C. W.: 1977, 'Tests of Competency to Consent to Treatment', *American Journal of Psychiatry* **134**, 279—284.
7. Roth, L. R., Appelbaum, P. S., Sallee, R., Reynolds, C. F. and Huber, G.: 1982, 'The Dilemma of Denial in the Assessment of Competency to Refuse Treatment', *American Journal of Psychiatry* **139**, 910—913.
8. Shelp, E. E. and Perl, M.: 1985, 'Denial in Clinical Medicine', *Archives of Internal Medicine* **145**, 697—699.

EDWIN R. DUBOSE, JR. AND EARL E. SHELP

COMPETENCY JUDGMENTS: CASE STUDIES IN MORAL PERSPECTIVE

I. INTRODUCTION

The value that we hold for others is shown, in part, in the ways that we express concern and respect for them. Our concern is revealed in the manner in which we care for their health, safety, and general welfare. Our respect, on the other hand, is for others as persons, with their own unique conceptions of themselves, of what gives their lives meaning, and their goals. If we are to respect the personhood and autonomy of others, we give priority to what they value and want, not to what we value and want for them or think that they ought to want. There obviously is possible tension between these two duties to promote another person's welfare and to respect his wishes [26]. Our society functions with an assumption that people have the capacity for free choice. Thus, we assume, until shown otherwise, that a person knows his or her mind better than we, and is, therefore, entitled to act on his or her reasoning, even if the action reflects a 'poor' choice based on preferences and choices that manifestly are not conducive to their health, safety, or welfare ([15], p. 2). Balancing the welfare of others with respect for their autonomy can be difficult in health care situations where a patient's competency is questioned.

II. MORAL ISSUES IN MATTERS OF COMPETENCY

The question of when a patient is incompetent is a troubling one ([12]; [2], pp. 117—131). Our society has a serious investment in the individual's capacity and right to define his or her own goals and to make choices designed to achieve those goals ([2], pp. 45—51). This interest in a respect for the freedom of persons (that one should regard others as rightly self-governing) is basic to moral life. However, autonomy is not absolute. The moral principle of beneficence is held to be another good in our society ([3], p. 148). If someone is incapacitated or if her misunderstanding of a situation is placing her at risk, in order to provide benefits and prevent or remove harms to others, we may be called to act in the person's best interests. The call to respect the freedom of others and the call to prevent harm to others raises the problem of ranking sometimes competing moral goods.[1] In the health care context, where the patient is vulnerable to the suggestions of physicians, this conflict can develop when the value systems of patient and physician do not correspond.[2]

Questions of competency in treatment contexts reflect and illustrate the conflict that care-providers often perceive regarding their moral obligations to patients [8]. The moral principles of beneficence and respect for the autonomy of persons are particularly relevant to this perceived conflict. It is a common perception among physicians that the patient has placed himself or herself in the hands of the care-provider, who by virtue of training and experience is in a position to make dispassionate and rational judgments in the patient's best interests. However, this therapeutic imperative to act in the patient's best interest often views that interest only from the perspective of 'medicine' or the perspective of the physician, perhaps in disregard of the patient's perspective and expressed wishes ([26], pp. 199, 218).

Clearly, there are times when the patient may be unable to be self-determining (as with dementia or an emergency situation in which the stakes are high and time is short), and the doctor or some surrogate must act in the patient's best interest, however defined. However, the competing principle of autonomy calls upon the physician to take seriously the patient's notion of what is good and attempt to work within that framework while maintaining a concern with the limits of good professional care. The danger is that the physician may absolutize certain values and project them upon the patient. Under such conditions, the patient becomes a victim of a medical tyranny as self-determination and participation in treatment decisions are eroded or entirely taken away.

There is a presumption by clinicians, as well as laypeople, that mental impairments, regardless of type or severity, justify informal (i.e., not authorized by a judge) determinations of incapacity and incompetency. This assumption, however, can be questioned. There are situations in which a person's capacity for judgment may be impaired but not altogether absent. Common experience teaches that people with certain mental deficits may be quite competent to make particular judgments or undertake certain activities while being incompetent to pursue other tasks or make other decisions. For example, a person judged incompetent to drive an automobile may not be incompetent to decide to participate in medical research, or may be able to handle simple affairs easily, while faltering before complex ones ([17], [18], [19]).

Other clinical realities, such as fear, stress, pain, or the effects of drugs, undermine or diminish, but do not obliterate, a given person's cognitive and volitional ability to perform the task required of them, *viz.*, to make their own decisions about medical care. As Perl points out, these factors appear outside the medical context, and, *per se*, are not indications of suspect reasoning ability in the ordinary course of human affairs. Also, some patients may have a flawed understanding of their situations due to the presence of denial, false beliefs, or a lack of clarity on the part of the physician ([10], [16], [28]).

There also is a presumption among laypeople and medical professionals alike that the concepts of capacity (a person's ability to perform certain tasks) and competency (the legal equivalent to capacity) have standard definitions and that uniform criteria are agreed upon to determine the presence or absence of capacity or competency. It is the case, however, that both concepts, especially competency, are subject to vigorous debates among philosophers and legal scholars with regard to their definition and the means by which capacity and competency are adequately assessed. As Perl indicates, there are no universal standards for competency.[3] A possible result of this state of affairs is not only intellectual and conceptual confusion, but also, as might be expected, inconsistent behavior by medical professionals, relatives, or surrogates of patients, and judges.

The lack of standard definitions and criteria for competency creates an atmosphere for medical paternalism in which the autonomy of the patient intentionally is limited by the physician who justifies coercion on the grounds of beneficence. However, there are many instances in health care in which the wishes of the patient simply do not match the judgment of the physician (as Perl points out, the issue of competence seldom arises as long as the patient agrees with the doctor). The issue is not so much who makes a decision (although that certainly is a concern), but that a decision is made that balances as fairly as possible the competing desires and interests of the parties involved. Health care professionals ought to be careful not to act in a coercive manner and subvert the patient's autonomy under the guise of beneficence. This undermining of the patient's self-esteem and autonomy can threaten the balance of rights and obligations that optimally characterizes the fabric of the health care relationship.

By defining competence as "a capacity to act voluntarily with sufficient information", Boisaubin recognizes the requirement to provide sufficient information to a patient who is able to understand that information and act upon it autonomously. If a situation is present in which there is insufficient information provided, or if the patient is unable for some identifiable reason adequately to process this information and act on it in a voluntary manner, Boisaubin advocates a course of action designed to eliminate the condition giving rise to the competency question. However, Boisaubin's approach or solution in several instances seems misdirected.

CASE I

In Case I, Boisaubin argues that "physiologic incompetence" may result from a metabolic problem that could impair the patient's decision-making ability. Thus, if "factors of organicity" are present, the doctor is in a stronger position to consider and justify an appeal to patient incompetence. However, in this case, these indicators are not present. The physician feels strongly that caution should be exercised. Since he feels that fear, anger, and denial of the possible severity of the patient's condition may be factors in that person's decision to leave the hospital, he proposes a period of hospitalization during which the patient could reassess his position. This time would be a 'grace' period enabling the patient to 'return' to a position from which he can more clearly participate in decisions regarding treatment. Boisaubin moves toward a 'weak' paternalistic strategy, the expressed reason for which is to enable the patient to recover rational equilibrium (the assumption being that this equilibrium is in some way impaired).

However, as Perl points out, these emotions are the patient's prerogative and do not necessarily indicate incompetence. The patient in Case I certainly finds himself in an abnormal, even frightening, situation, but this condition in itself does not support the presumption that he cannot function in an adequate manner. The doctor must be careful not to superimpose his or her wishes. As Boisaubin states and Perl concurs, the purpose is not to make the patient do what the doctor believes that he should, i.e., that the patient is competent only if the patient agrees with the physician. From the physician's standpoint, a patient may make an 'unwise' choice, but that choice in itself does not imply that the person is incompetent.

Boisaubin also seems to propose that the physician act in a strongly paternalistic manner by altruistically manipulating the patient into staying. The physician represents a powerful social role in our society and, consequently, possesses authority. There is a concomitant social and moral duty to use this authority responsibly. Many patients do not realize that they have the right to disagree with the doctor. It may be too easy to use the weight of the office to give the patient a 'hidden' message of censure, to intimidate the person into remaining hospitalized, or risk a confrontation with the doctor's authority. If he or she does resist the physician's manipulation, there is still the risk that the patient will not return for subsequent care.

Boisaubin recognizes the 'sin of coercion' in a relationship characterized by the tension between autonomy and beneficence. Coercion is difficult to justify, and should be reserved for special circumstances or when the medical condition clearly warrants it. If indications are such that the doctor chooses to err on the side of caution, he or she can attempt to impress the patient with these concerns. This impression would carry more weight if a respect for the patient's autonomy is the basis for the therapeutic relationship.

CASE II

In the second case, Boisaubin makes an argument for a weak paternalism in illustrating how a decision can be reached on the issue of competency. Although the underlying terminal condition can only be palliated, the dilemma for the physician is that the possible harms of the removal of the endotracheal tube may outweigh the benefits of the tube's removal. Since the situation raises the issue of possibly hastening death, the physician would want to consider the patient's request carefully, as Bousaubin does. Boisaubin is concerned that the patient, faced with a frightening condition and unsettling ICU environment, may not fully understand the implications of his request. He proposes a 24-hour period for the patient to consider his decision, to discuss the procedure, as much as possible, with family and friends, giving the patient the opportunity to change his mind.

On the face of them, Boisaubin's suggestions seem reasonable. The ICU is not a normal setting, and the patient's situation is serious. Boisaubin seems to think that this unfamiliar environment and weakened physical condition can lead to fear, confusion, and anxiety which may 'impair' the patient's competency. Conferring with loved ones does provide a chance for additional reflection. However, one implication of this type of approach is that the patient is thought not to know his own best interest, that the concern for incompetence is strong enough to warrant a paternalistic request for time and reflection so that the patient will recover a 'reasonable' point of view by changing his mind about removal of the tube.

However, the 24-hour period is quite arbitrary. If the interest on the part of the physician is to gain confidence that the patient is clear about what he is requesting, why not settle on 48 or 72 hours? Sometimes there is a tendency on the part of doctors to relieve their own concerns while claiming to act in the patient's best interests, respectful of his or her autonomy. The discomfort of the tube, of trying to communicate by writing, and the confusing atmosphere of the ICU with its task orientation and sometimes curt staff may contribute to the patient's desire to forego or withdraw support, but this delaying strategy may indicate a desire of staff to have the patient do what is deemed 'reasonable' (or least troubling to them), not necessarily a concern to 'respect' the patient's wishes.

The terminal nature of the patient's condition may be a factor as well. As Perl says, the decision to withhold or withdraw potentially lifesaving treatment should not be made lightly. However, courts have held that a competent patient has the right to withhold consent and refuse any life-saving treatment ([30], [22]). It would seem that the person with inoperable lung cancer has the right to request withdrawal of treatment even if this action would hasten death. The patient in Case II is not clearly incompetent. Certainly, this is a scenario that calls for reflection. If after the physician has carefully explained the possible consequences of his request and the patient maintains his desire, then the clinician is morally obliged to honor it.

CASE III

The confusion over definitions and criteria of competency for patients with physical impairments is also possible for patients with psychiatric and neurologic impairments ([29], [5], [14]). In Case III, Boisaubin is correct to consider schizophrenic delusion, misinterpretation of medical reality (possibly intensified by fear of radiation), and the possibility of metabolic distortions of judgment in evaluating the woman's competency to refuse radioactive therapy for her hyperthyroidism. However, the fact that a patient is mentally impaired, or unable to perform some other function competently, while highly relevant, is not necessarily determinative of competency in medical decision-making.

Courts have criticized physicians for imputing incompetency simply because a patient's decision is not a medically reasonable one [17]. Too often the term 'irrational' (meaning incompetent) is used in a pejorative rather than an analytical sense. In these instances, resistance to the physician's desire is termed 'irrational'.[4] It may be that the patient's reasoning does not conform to the doctor's expectations, but refusal is not necessarily a sign of incompetence.

Persons committed to mental institutions have been found to be competent to decide about their medical care because their mental impairment did not interfere with their ability to understand the nature and consequence of the proposed treatment ([23], pp. 40–47). Whether in a general medical or a psychiatric context, it is clear that a competency determination must be directed at the specific decision at hand. As Boisaubin argues, if the first three possibilities are eliminated, this patient must be allowed the opportunity to make an informed decision, even if it is based upon the patient's unique set of values. The physician, as 'gatekeeper' of therapeutic options, has the responsibility to educate the person about treatment options by providing sufficient information, to assist her in making her decision about therapy, and to respect her right to self-determination, provided that she is capable of giving informed consent.

CASE IV

In Case IV, the patient desires a mode of therapy that the doctor views as unconventional and ineffective. Boisaubin reminds us that the physician has the duty to disclose, to extend information, to explore assumptions, and to offer a clear explanation of the condition, possibilities, and risks of the therapy. If the physician has done this, he or she must respect the patient's informed decision even if it appears 'unwise'.

A physician has an obligation to the profession. If a physician feels that the patient's decision will be harmful to the person's well-being, if he or she is uncomfortable with the patient's actions, the physician may withdraw from the case. Since some relationship has been established, the doctor has the moral obligation to remain a sustaining presence until the patient has found another physician.[5] However, since the profession is dedicated to the well-being of the patient (and must, therefore, appreciate the moral principle of autonomy), Boisaubin rightly suggests possible therapeutic negotiations and compromises in which the course of treatment least objectionable or most agreeable to the patient (and the physician) is followed. Such a compromise arrangement, if medically feasible, would serve to minimize the patient's distress and augment her role as co-determiner of treatment, while still providing an acceptable (though perhaps not optimal) treatment option in which the physician can participate with a clear (or adequate) conscience.

III. CONCLUSION

These cases illustrate that standards of competency are evolving and, therefore, treatment options in situations of questioned capacity must remain flexible. As moral principles, autonomy and beneficence may appear to or actually conflict in the medical relationship. Too often, however, these principles are set in opposition to each other, creating a struggle for control over who decides, obscuring the issues of how and what decisions are reached ([4], pp. 105–107). The moral issues involved in competency questions are not necessarily either/or in nature; autonomy and beneficence are not necessarily polar opposites. The creation of such a dichotomy itself has a moral dimension since the tendency to simplify complex issues easily turns right into good, wrong to bad, and creates the issue of paternalism.[6] Since determinations of competency can have legal as well as clinical implications, clarity and consistency should be objectives in the application of competency judgments. Unless certain conditions apply, the authority to make decisions lies with the patient, and the physician is called to support and augment the patient's choices. It is the burden of the physician to justify overriding patient autonomy in the name of beneficence. Yet if the scope of analysis can be broadened to show the fallacy of a perceived dichotomy, we can promote caring *and* mutual respect in the therapeutic relationship, a relationship the form and content of which acknowledges the plurality of sometimes competing principles and tries to draw upon the strengths of each.

The Park Ridge Center for the Study of Health, Faith, and Ethics
Chicago, Illinois

Foundation for Interfaith Research and Ministry
Houston, Texas

NOTES

[1] For a discussion of the diversity within the moral community, see [25], pp. 7—25. For a discussion of the difficulty of establishing the objectivity of any moral preference, see [9], pp. 32—37.

[2] For a broad discussion of the principles of beneficence and autonomy and their application to health care, see [27].

[3] There have been diverse attempts to define competence and establish criteria within a general medical model on which a clinical judgment can be based. Roth, Meisel, and Lidz discuss five categories for competence; see [24]. Appelbaum and Roth suggest four possible standards for judging competency in [1]. Other scholars argue for a single standard; see [20], [21]. Perhaps because of the difficulties in establishing a common definition and set of criteria for competency, several scholars advocate a sliding scale or continuum; see [3], p. 70; [6], pp. 106—107.

[4] For a discussion of one ideological component of the physician's view of patient's inability to understand medical matters, see [11], pp. 184—185.

[5] For a discussion of 'sustaining presence', see [25], pp. xi, 93—98, 174, 204—205.

[6] For a discussion of the problem of perceived dichotomies, see [13]. In applying Gould's argument to competency issues, too strict a concern with autonomy versus paternalism will create a struggle for control within the medical relationship, may lessen the chances for patient compliance with treatment, and may increase the tendency of clinicians to interpret competency as agreement with what the physician desires. Also, see [26], p. 199.

BIBLIOGRAPHY

1. Appelbaum, P. and Roth, L.: 1982, 'Competency to Consent to Research', *Archives of General Psychiatry* **39**, 951—958.
2. Beauchamp, T. and McCullough, L.: 1984, *Medical Ethics*, Prentice-Hall, New Jersey.
3. Beauchamp, T. and Childress, J.: 1983, *Principles of Biomedical Ethics*, Oxford University Press, New York.
4. Brody, H.: 1981, *Ethical Decisions in Medicine*, Little, Brown and Company, Boston, Massachusetts.
5. Brown, P.: 1986, 'Psychiatric Treatment Refusal, Patient Competence, and Informed Consent', *International Journal of Law and Psychiatry* **8**, 83—94.
6. Childress, J.: 1982, *Who Should Decide?* Oxford University Press, New York.
7. Drane, J.: 1975, 'The Many Faces of Competency', *The Hastings Center Report* (April), 17—21.
8. Dworkin, G.: 1977, 'Paternalism', in S. J. Reiser, *et al.* (eds.), *Ethics in Medicine*, MIT Press, Cambridge, Massachusetts, pp. 190—198.
9. Engelhardt, H. T., Jr.: 1986, *The Foundations of Bioethics*, Oxford University Press, New York.
10. Faden, R. and Beauchamp, T.: 1980, 'Decision-making and Informed Consent', *Social Indicators Research* **7**, 313—336.

11. Freidson, E.: 1961, *Patients' Views of Medical Practice*, University of Chicago Press, Chicago.
12. Gert, B. and Culver, C.: 1981, 'Competence to Consent: A Philosophical Overview', in *Competency and Informed Consent*, N. Reatig (ed.), National Institutes of Mental Health, Rockville, Maryland, pp. 12—31.
13. Gould, S. J.: 1986, 'Archetype and Adaption', *Natural History* **95** (October), 16—27.
14. Grossman, L. and Summers, F.: 1980, 'A Study of the Capacity of Schizophrenic Patients to Give Informed Consent', *Hospital and Community Psychiatry* **31**, 205—207.
15. Harris, J.: 1985, *The Value of Life*, Routledge and Kegan Paul, London.
16. Knight, J.: 1977, 'Judging Competence: When the Psychiatrist Need, or Need Not, Be Involved', *The Hastings Center Report* **7**, 19—20.
17. Lane v. Candura 376 N.E. 2d 1232, 1235 (Mass. App., 1978).
18. Meisel, A. and Roth, L.: 1981, 'What We Do and Do Not Know About Informed Consent', *Journal of the American Medical Association* **246** (November 27), 2473—2477.
19. Meisel, A., Roth, L. and Lidz, C.: 1977, 'Tests of Competency to Consent to Treatment', *American Journal of Psychiatry* **134** (March), 279—284.
20. Olin, G. A. and Olin, H. S.: 1975, 'Informed Consent in Voluntary Mental Hospital Admission', *American Journal of Psychiatry* **47**, 938—941.
21. Owen, H.: 1977, 'When is a Voluntary Commitment Really Voluntary?', *American Journal of Psychiatry* **47** 104—110.
22. Quackenbush, 156 N. J. Super 282, 383 A. 2d 785 (1978).
23. Robertson, J.: 1983, *The Rights of the Critically Ill*, Ballinger, Cambridge.
24. Roth, L. et al.: 1977, 'Towards a Model of the Legal Doctrine of Informed Consent', *American Journal of Psychiatry* **134** (March), 285—289.
25. Shelp, E. E.: 1986, *Born to Die?: Deciding the Fate of Critically Ill Newborns*, The Free Press, New York.
26. Shelp, E. E.: 1982, 'To Benefit and Respect Persons: A Challenge for Beneficence in Health Care', in E. E. Shelp (ed.), *Beneficence and Health Care*, D. Reidel, Dordrecht, Holland, pp. 199—222.
27. Shelp, E. E.: 1982, *Beneficence and Health Care*, D. Reidel, Dordrecht, Holland.
28. Soule, R.: 1979, 'The Case Against Total Candor', *Medical World News* **20** (May 14), 94.
29. Stanley, B. and Stanley, M.: 1982, 'Testing Competency in Psychiatric Patients', *IRB* (October), 1.
30. In re Yetter, 62 PA.D. and C.2d 619 (1973).

PART V

COMMENTARY AND CRITIQUE:
ANOTHER LOOK AT CONCEPTS
OF COMPETENCE

HAROLD Y. VANDERPOOL

THE COMPETENCY OF DEFINITIONS OF COMPETENCY*

Each of the essays in this volume contributes to the literature on definitions and determinations of patient competency, and all demonstrate that there are essential and inherent interconnections among philosophical distinctions, patterns of ethical reasoning, legal traditions, and clinical decision-making. After noting how several of the authors of these essays define competency and after flagging some of the notable contributions of these authors, this paper will identify and briefly discuss two unexplored areas of concern. Namely, it will indicate how the world-views of patients and the temperaments of patients have a critical bearing on when they are considered competent, incompetent, or semi-competent.

Beauchamp defines patient competency as a general ability to understand and weigh relevant information regarding medical risks and benefits and to make a decision to accept or reject treatment in light of a person's "relatively stable values" ([2], pp. 51—55). He says that in order to exercise competency within medical settings, patients must be autonomous. Beauchamp's minimalist account of patient autonomy means that patients can respond with conventional resistance to ordinary social pressures and influences ([2], pp. 63—64). He argues convincingly that any selection of competency tests, as well as any determination of competency thresholds, necessarily rests upon moral criteria, the most critical of which are the 'essentially contested concepts' of respect for the individual's autonomy and benevolent paternalism ([2], pp. 66—67).

Knight [9] surveys the contents and limits of several definitions as he displays how judgments about competency are exceedingly complex and many faceted. His richly-faceted account of these judgments considers patients' cognitive abilities regarding understanding, reasoning, and choosing ([9], pp. 17—19), whether they consent to or refuse treatment ([9], p. 13), what the risk-benefit ratios of treatments are ([9], p. 13), whether patients are retarded, chronically institutionalized, or suffering from severe mental illnesses ([9], pp. 20—21), whether patients are cognitively and effectively aware of their life circumstances

197

Mary Ann Gardell Cutter and Earl E. Shelp (eds.), Competency, 197—210.
© 1991 *Kluwer Academic Publishers. Printed in the Netherlands.*

and can function in their respective environments ([9], pp. 19—22), whether they can pass a mental status examination regarding orientation, memory, mood swings, and rational impairments (especially delusions and hallucinations) ([9], pp. 5, 10, 21), and whether 'spurious indicators' like anger toward medical professionals or hospitalization are causing patients to act or react in unusual ways ([9], pp. 10—11). Knight furthermore provides a useful case study that illustrates various aspects of competency determinations ([9], pp. 6—10), and he notes how and why disagreements over competence arise between psychiatrists and other medical specialists ([9], pp. 6—15). He stresses that the rationality of the patient's choice must be assessed according to a nonjudgmental understanding of the patient's own framework of meaning or "belief system" ([9], pp. 9—10, 18—19).

After surveying the value of automony in law and ethics, Robertson ([14], pp. 127—128) defines competency as the capacity to perform the task of exercising autonomy and says that, within the law, there is a consensus that a patient's competency to make medical decisions is compatible with civil commitment, intermittent lucidity or disorientation, and even irrational behavior ([14], p. 132). Robertson points to several medical contexts in which decisions regarding competency are raised ([14], pp. 128—130) and argues that on legal and ethical grounds competency determinations should be viewed not as a continental dividing line but as one point on a map of considerations, including whether to treat and who makes treatment decisions ([14], pp. 143—144).

Pellegrino defines competency as the ability to make a reasoned choice from alternative courses of medical treatment. This includes the tasks of receiving, comprehending, weighing, integrating, and choosing; and these abilities need not be global or absolute, but should be "substantial" ([13], pp. 29—32). Pellegrino focuses on physiological impediments to competency, which impediments the doctor is warranted in removing for severely or acutely ill patients even if they give prior instructions opposing such treatments. Such action is warranted, says Pellegrino, because it serves both beneficence and patient autonomy ([13], pp. 34 and 38). He provides helpful perspectives both on impediments (like time constraints) to a physician's making reasoned choices, and he accents the importance of making sure that each patient's proxy is competent and free from conflicts of interest ([13], pp. 32—33, 36—37). The patient's rationality should be assessed not in terms of the doctor's criteria regarding logic and rationality, but in light of the internal and logical consistency of the patient's own 'value system' or presumptions ([13], pp. 33, 36).

Like Pellegrino, Morreim restricts her discussion to seriously ill patients. Morreim defines competency as the capacity to make autonomous decisions regarding medical care, and says that these decisions should be viewed as *whatever* persons voluntarily and intentionally wish to do in light of their "core reasons for thinking, choosing, and acting" ([12], pp. 99—100). Autonomous persons organize their lives and make choices "according to a more-or-less coherent picture of the world", which provides "some stability" for values and beliefs over time ([12], pp. 100, 106). Great tolerance should be exercised toward those who voice preferences or make inferences out of religious frameworks of meaning ([12], p. 105). Since patients should have the greatest possible control over their lives ([12], p. 111), their competency should be assessed not in terms of definitive principles and tests, but in light of suspicious signs of rational impairment, namely, when they voice beliefs that are "obviously false", "flagrantly unreasonable", or widely at variance with their past beliefs and values ([12], pp. 103—107). Compared to the courts and the working definitions of lawyers, doctors are much better suited to deal with "those difficult cases" of persons who are partially impaired or who manifest dubious competence" ([12], p. 116).

As indicated in several of the definitions and quotations just cited, definitions of competency usually appeal to the world views or belief systems or patients. These world views (used interchangeably with belief systems or frameworks of meaning) are regarded as essential for self-identity and autonomy, for they provide persons with a more or less stable, coherent, and internally-consistent understanding of one's place in society and the world generally. World views or belief systems serve as an essential ingredient in competency determinations because they are viewed as defining and determining boundaries or thresholds of acceptable rationality. Even if some particular world view appears to be non-rational or even irrational, the choices of patients based on that world view must be respected. Definitions and discussions of competence thus commonly assume that the rationality, reasonableness, or logical sophistication of patients need only reflect the internal 'logic' of some world view and the 'reasons' (that is, explanations or explications) inherent to it ([13], pp. 33—36; [9], pp. 10—11, 18—19; [12], pp. 96—97, 100—106).

This commonly-voiced perspective regarding world views appears to serve at least two functions for those who hold to it. First and quite obviously, it fosters tolerance and diversity regarding the degrees of rationality required of patients — tolerance compatible with first amendment provisions regarding religious freedoms in the U.S. Constitution. Second, appeals to a patient's world view are regarded as providing a standard by which eccentric patterns of shared beliefs are distinguishable from patterns that are indicative of mental instability or insanity.

Although one is tempted to leave most things supportive of tolerance and autonomy alone, one can nevertheless wonder what these references to world views, frameworks, or belief systems, amount to. Is this short-hand language for a Christian Scientist, a Jehovah's Witness, or some well-known religious denomination? To what degree are world views actually stable or internally consistent? What constitutes a stable, not to speak of a more-or-less stable, picture of the world? Is there some acceptable and/or unacceptable time frame for a world-view that is stable versus one that is not? Must a more or less sizable social group believe in this framework of meaning before it can be considered stable? And what are the criteria for an internally consistent or coherent belief system? Must it have a cosmology? A particular view of human nature or the natural world? A particular view of the human body or of an afterlife (or lack of one)? Must it have all of these? Must they be logically related? How many aspects of life must a 'framework of meaning' address? Do such frameworks provide us with threshold definitions of logic and rationality?

Criteria like stability and coherence as applied to some world view — and *ipso facto* to acceptable levels of rationality — are so vague that one wonders if they can be normative — as discussions of competency presuppose. Assuming that they are normative, presumably the refusal of a relatively low-risk, high-benefit medical treatment by a Fire Baptized Holiness Christian would be honored as a competent decision if the patient (1) passes certain mental tests, (2) does not have his autonomy compromised by pain, illness, and so on, and the medical team discovers (3) that a group that calls itself by this name exists and is stable, (4) that many of its beliefs are relatively consistent with one another, (5) that these beliefs could readily support such a refusal, and (6) that this patient either identifies with or is a member of the group.

When do medical professionals ever pursue points (3) through (6)? Why are these points necessary; and if they are, how could they be examined or tested? Who could conduct such an examination competently? Assuming that they could, should, and would be examined, how would one decide whether the world view of Fire Baptized Holiness Christians generally and this patient's particular understanding of that world view satisfies criteria of stability and internal consistency? Would a discovery that the group with which this patient identifies is a patently unstable, store-front church affect his being considered competent? What would count for this group's representing mass delusion or mass hysteria (whatever these are)? If the group with which this patient identifies is found to be socially and emotionally 'stable', what should be done to test the emotional stability of this patient?

Several of these questions and comments may appear somewhat outlandish. If so, criteria like stability and internal consistency are probably not being taken seriously — at least not as seriously as the above face-value probing of these presupposes. Serving as they do, however, as fundamental criteria for a patient's rationality, and hence as a critical threshold in competency decisions, issues pertaining to the definitions and functions of world views deserve serious exploration. Since world views serve as minimalist and tolerant definitions of rationality, much remains to be investigated about them. As points of departure, the criteria inherent in definitions of world views or social constructions of reality need to be investigated ([5], pp. 87—125). These need to be related to questions of logic and fact, to the belief 'systems' of many types of social groups, and to notions of mental illness (including mass mental illness). The expected ethical and policy-related functions of these definitions and criteria need to be clarified. And all of these topics need to be tested by actual cases.

A second set of issues that bears on determinations of competency and that needs further scrutiny involves the temperaments of patients. Ordinary language contains numerous labels and distinctions of various human emotional states. Some of these refer to temperament*s*, that is, to human dispositions, states of being, or emotional make-ups. Persons are thus perceived as being frivolous rather than sober or serious; as demonstrative rather than stoical or restrained; as impulsive rather than deliberate; as capricious rather than steady or predictable; as flexible rather than rigid; as shy or reserved rather than outgoing or assertive; as aimless rather than purposeful or directed; as idealistic rather than realistic; as tenuous or indecisive rather than decisive or strong; as impassive or stolid rather than responsive or sensitive, and so on.

Many of these distinctions of temperament, along with a number of additional ones, are often applied to humans not in terms of some characteristic emotional constitution, but as tempera*mentalness*, that is, as time-limited and/or context-limited emotional states — as moods or inclinations. Humans are thus assessed as frivolous, shy, aimless, or tenous now or in some given context, not as generally or characteristically frivolous, shy, or aimless. Some distinctions seem inherently to refer to such time-bound and/or context-limited emotional states — for example, when persons are characterized as effusive, wishy-washy, silly, irritable, or brusque.

There is a sense in which the temperaments and temperamentalness of patients may have little bearing on decisions regarding their psychological and legal competency. Most persons who are perceived characteristically or for the time being as stoical, or idealistic, or shy, or tenuous can likely pass mental tests regarding comprehending and weighing medical harms and benefits, then choosing whether to accept or reject medical care. They thus can be construed as stoical or idealistic, yet also competent. They can even characteristically act and think indecisively, yet be able eventually (perhaps with the aid of much counsel and urging) to select some course of action. In such instances an open-question test could be applied: "This person's temperament is X, but is she competent?" The answer could often be "Yes" in terms of some competency threshold, that is, competency as distinguished from psychological or legal incompetency.

Insofar as competency is distinguished from semi-competency or partial competency, however, the temperaments of patients have a direct bearing on judgments of competency. This distinction is by and large neglected in discussions of patient competency and likely gives rise to a host of significant issues, some of which will now be broached.

As a beginning point, it would appear that several of the characteristic temperaments just identified run counter to certain dominant ideals and assumptions regarding the *emotional* dynamics of good decision-making regarding medical procedures, namely, that medical information, possible harms and benefits, and so on, need to be thought about seriously, deliberately, realistically, and decisively. Even tests of competency seem inherently to assume the normative status of these ideal moods or temperaments. As evident in several of the papers in this collection, tests for competency include the patient's ability to understand disclosed information, to weigh the risks and benefits of alternative treatments, and to be able to choose some course of action. Given the ethos of medical care, however, these tasks become code words for something like being serious and realistic with the information presented, deliberating over various options, and evincing some degree of decisiveness. In other words, doing the tasks associated with competency seems inherently associated with normative, ideal states of being and feeling. They represent ideals insofar as medical professionals need not and often do not manifest only these moods and temperaments when they make decisions, but these moods and temperaments can be and are evoked as standards required of patients.

In light of idealized and normative temperaments within medical settings, the characteristic temperaments of many patients are decidedly out of place. To put it mildly, persons risk being considered strange if they are frivolous, impulsive, capricious, aimless, idealistic, or impassive in situations that are (1) assumed to be utterly serious, realistic, and endangering and (2) when these situations are regarded as resolvable or partially resolvable by relatively deliberate, decisive, and purposeful thought, as well as by a willingness to share thoughts and feelings with others. Indeed, persons who cannot adjust their temperaments to these implicitly normative temperaments risk being viewed — on a sliding scale — as idiosyncratic, bizarre, queer, aberrant, abnormal, emotionally disturbed, or possibly insane.

Of course, frivolity in the face of dangerous or threatening circumstances may be indicative of a mechanism of defense or even a form of mental illness. But then again, it may not signal these. Levity in the face of serious circumstances could represent a common, valued, and chosen way of being when confronted with a drill sergeant, an English teacher steadfastly committed to clarifying distinctions between direct objects and predicate nominatives, or a doctor who exudes the characteristic ethos of his or her medical speciality. Persons like Luke in *Cool Hand Luke* may be idiosyncratic and strange for a great many reasons without necessarily being mentally imbalanced. If this holds for frivolity, it is all the more applicable for persons who are impulsive, impassive, and so on. In short, the underlying moods, emotions, and value assumptions that commonly accompany terms like understanding, weighing, and choosing when these are utilized as normative criteria in medical care settings need to be explored.

Interesting perspectives on how the temperaments of patients affect judgments regarding their semi- or partial competence are found in the literature of trait psychology and psychiatry. Within this literature temperaments are usually referred to as psychological traits, personality styles, or personality types. Trait psychologists have identified over 18,000 English words that refer to various personality variances (optimistic, pessimistic, idealistic, rigid, responsive, aloof, and so on). They typically cluster these variances into 20 or more personality styles, defend these as accurate and real (even if no more directly observable than most other psychological concepts like conditioning or ego or libidinal fixations), regularly explore how these styles or traits are developed, and, importantly, often distinguish between personality styles that are normal from those that are not [4].

Definitions and discussions of abnormal temperaments or personality styles display a wide and intriguing set of opinions regarding how persons are assessed and manipulated in clinical settings. Wanting both to encourage doctors to acknowledge their feelings about patients and to move beyond common labels such as complaining, or angry, or argumentative patients, John J. Schwab and Henry W. Brosin, in their book on psychiatric consultation, list a variety of types of persons with "personality disorders" and propose how they should be handled. Categories of specific disorders include the dependent and overdemanding personality; the orderly and controlled personality; the dramatizing, emotionally involved, and captivating personality; and the uninvolved and aloof personality ([15], pp. 151–152). Each of these categories is attached respectively to labels from "psychoanalytic models", namely, patients orally fixated, compulsive, hysterical, and schizoid. "Suggestions for appropriate management" of these patients include making only "minor concessions" to dependent and demanding patients, discussing the underlying "anxieties" of emotionally involved patients, and preventing the "complete withdrawal" of aloof patients from therapy.

The authors proceed to discuss more fully certain patients with "personality trait disturbances", those whom they say represent "the great majority of management problems", which often go undiagnosed. These include "passive dependent" personalities (those who appear helpless, childlike and resourceless when faced with life-threatening illness), "independent" personalities (those who deny their dependency, disregard admonitions from staff, and "insist on following their own inclinations"), and "compulsive" personalities (those who are rigid, excessively worried, and cause frictions with the staff when anxiety mounts) ([15], pp. 154—159). The authors remark that the psychiatric consultant "has no problem" identifying independent personalities because the "basic disorder which generates such a drastic defense" (guised as independence) is denial. They urge the consultant first to keep such patients from being restless or from signing out of the hospital, then to help them "maintain their self-esteem which has been jeopardized by the illness" and which may have been lost when their denial is "shattered" by the consulting psychiatrist or staff ([15], pp. 157—158).

Exemplifying a more recent study, Chase Patterson Kimball says that his personality groupings are neither "pejorative" nor "diagnostic", but soon remarks that these labels are "pragmatically useful for the clinician, from the standpoint of both diagnosis and therapy" ([8], pp. 27—29). He notes that all persons gravitate toward some given "neurotic pattern" of feeling and acting and that such patterns are not "pathological" unless they cause maladaptivity, disharmony, and a loss of inner tranquillity. The "personality styles" identified include obsessive-compulsive, hysterical, impulsive, depressive, and infantile personalities ([8], pp. 31—38). To focus on the last of these, infantile personalities are marked by arrested or faulty social learning, impaired self-expression, dependency, passivity, and over-extended ideals. Toward these patients the doctor needs to display "benevolent firmness", to point out how they manipulate others with their feelings and complaints, and to show how their complaints and organ symptoms are objectified expressions of their feelings ([8], p. 37).

'Personality Disorder' analysis appears to have achieved virtual orthodoxy by 1980 when the third edition of the *Diagnostic and Statistical Manual of Mental Disorders* (DSM-III) was published by the American Psychiatric Association ([1]; see also [6]). DSM-III instructs its users that personality disorders are "inflexible and maladaptive" personality traits, which traits are defined as "enduring patterns of perceiving, relating to, and thinking about the environment and oneself" ([1], p. 305). Four of the eleven specific personality disorders outlined are persons who are schizoid (who are cold, aloof, indifferent to praise or criticism, and form close friendships "with no more than one or two persons"), histrionic (those who are self-dramatizing, crave excitement, and are overreactive, irrational, and shallow), avoidant (those who are hypersensitive to rejection, humiliation, and shame, and are withdrawn, desirous of affection, and experience low self-esteem), and atypical, mixed, or other personalities (those who manifest features of several specific disorders or who are judged to be masochistic, impulsive, immature, or maladaptive in some other significant way) ([1], pp. 306—330).

These brief references to selected medical writings indicate that doctors are to turn a trained eye, rather than a blind eye, to the temperaments of their patients. Against the background assumption that temperaments are ubiquitous features of humanness, medical writers focus on distinguishing and characterizing disordered or maladaptive persons-personalities. Although its meanings are not altogether clear, 'disordered' implies more than idiosyncratic or unusual dispositions of character — dispositions that can and do evoke special degrees of curiosity, enjoyment, and respect from others. For the above authors, disordered implies important degrees of psychological impairment, social disruptiveness, and a type of orientation that brings pain upon oneself and others ([15], [8], [6], [7], [16]). While those with disordered or maladjusted personalities are distinguished from those manifesting psychopathology ([8], pp. 28—29; [11], pp. 108—109), some disordered types (like schizoids) are sometimes thought to display features precursive of certain forms of psychopathology (like schizophrenia) ([6], pp. 109 and 119—120; [10], pp. 224—226).

Beyond descriptions of the characteristic reactions, symptoms, and etiologies of disordered or maladaptive personality types [1], medical literature specifies two types of handling and treatment for persons diagnosed with these conditions. First, persons so diagnosed are viewed as subjects needing individual and group psychotherapy in order to assist them and those relating to them to understand and overcome some of the problems and pains they create ([8], pp. 31—38; [1], pp. 129—130; [6], pp. 125—130; [11], pp. 215—275). Second, and already indicated, persons labeled as disordered are viewed as creating special problems for themselves and medical professionals in health care settings ([15], pp. 150—161; [8], pp. 15—38; [6], pp. 104—130; [7], pp. 108—123; [10], pp. 224—230; [16]). This second body of literature raises issues directly and indirectly related to judgments regarding the full or attenuated competency of patients. A brief scrutiny of the management recommendations regarding the independent and hysterical personality types just mentioned will highlight some of these issues.

The independent personality disturbance or disorder identified by Schwab and Brosin typifies individuals who deny how dependent they are, disregard admonitions from medical staff, and insist on following their own inclinations. They are said to represent one type of disturbed personality because they drastically defend their independence, when in fact they are dependent. Their independence in the face of the knowledge and recommendations of medical professionals is thus viewed as rooted in a denial of reality. Although not psychopathic, they are psychologically disturbed and maladjusted. This warrants professional interventions that will get these patients to change their minds, and then enable them to adjust psychologically to the fact that their wishes were reversed and their orientations transformed.

Or consider the hysteric persons-personalities as defined by DSM-III and other sources. As dramatic, overly-emotional, over-reactive, shallow and often irrational, hysterical persons-personalities are said to lack a capacity "for intense or persistent intellectual activity", to be unable "to describe things with sharp definition and precision", and to react "in characteristic dramatic" ways when faced with physical difficulties ([8], p. 34). Hysterical patients should thus receive "reasoning guidance" and matter-of-fact certainty from the doctor. Hysterics manifestly "do not require precise and detailed descriptions of their illness", for this will only make them more uncomfortable, sometimes extremely so. The doctor should thus confront the patient's denial because this denial can "interfere with diagnostic procedures or therapy". Doctors may also have to utilize "ingenious efforts" to convince such patients which courses of action to follow ([8], p. 34).

These assessments assume that these patients reason, understand, weigh, and choose diagnostic and treatment alternatives in atypical, faulty, inept, and insufficient ways. The information that they receive, the weighing they do, and the choices they voice must therefore be influenced, guided, supplemented, revised, discounted, or reversed by medical professionals. Similar assessments are made for other 'disordered' personality types, including avoidant, impulsive, borderline, suspicious, antisocial, or schizotypical persons.

Interestingly, none of the authors cited here directly poses the question as to whether individuals with these types of personalities or characteristic temperaments are competent as opposed to psychological or legal imcompetency — because it is apparently assumed that these patients are competent compared to persons who are, for example, psychotic, hallucinating, or seriously impaired by an organic brain syndrome ([10], pp. 226—227). Nevertheless, management recommendations for these populations of patients assume that although not clearly irrational, their thinking processes are not truly rational, that although they can weigh information, they do so in disordered and problematic ways, and that although they express choices, these choices often or typically reflect vexing psychological states like denial and fear. In short, according to minimalist legal and functional definitions of incompetency, they seem competent, but when assessed according to underlying normative notions of mental health utilized by psychologists and psychiatrists, they are, at best, semi-competent. They are incompetent in the sense that they cannot be trusted to make decisions effecting their lives without the interventions of medical professionals.

This altogether brief analysis underscores some of the ways the temperaments of patients bear importantly on judgments regarding their competency and autonomy, specifically their emotional competency to exercise their autonomy. This analysis calls for critical and sustained scrutiny of medical assumptions and traditions regarding persons viewed as less than fully competent. The agenda of issues that bear scrutiny includes an analysis of the epistemological status of psychological types or temperaments (including the relationships of these to identifiable emotional reactions of ethnic and religious groups) ([3], pp. 153–158); an understanding of how these are informed by concepts of normality or mental healthiness and abnormality or mental disease; an investigation of the degrees to which understanding and managing these persons are shaped by traditional ideals and role-expectations of doctor and patients respectively (literature on doctors who become patients could inform such studies); an assessment of how the insights and conclusions of these studies does or does not call for differing definitions of concepts like rationality, understanding, weighing, and choosing; and further clarifications regarding policies supportive of psychotherapy for these persons and policies pertaining to their management and decision-making in the face of illness and disease in the hospital. Even if doctors are "much better suited" to address "difficult cases of partically impaired or dubious competence" than judges and lawyers ([12], pp. 26–27), the assumptions, roles, and policies of doctors deserve much more critical scrutiny.

Each of the broader subject areas identified here — the world views and the temperaments of patients — further highlight the value of interdisciplinary analysis. In addition to philosophy, ethics, law, and clinical medicine, however, investigations of the world views and temperaments of patients can also profit greatly from the analytical insights of other disciplines of inquiry, notably language, literature, religious studies, and psychology — as is partly evident in this analysis. The competency of definitions of competency thus depends upon a complete capturing and displaying of the factors at play in such definitions. This completeness cannot be achieved unless respective disciplines of inquiry truly inform and partially transform one another.

University of Texas Medical Branch
Galveston, Texas.

NOTE

* Parts of this paper are indebted to the critical comments and helpful suggestions of William J. Winslade, Thomas H. Murray, and Nancy Rhoden, to whom I offer my special thanks.

BIBLIOGRAPHY

1. American Psychiatric Association: 1980, *Diagnostic and Statistical Manual of Mental Disorders*, third edition, American Psychiatric Association, Washington, D.C.
2. Beauchamp, T. L.: 1991, 'Competence', in this volume, pp. 49—77.
3. Bernstein, L. and Bernstein, R. S.: 1980, *Interviewing: A Guide for Health Professionals*, third edition, Appleton-Century-Crofts, New York.
4. Endler, N. S. and Hunt, J. M.: 1983, *Personality and Behavioral Disorders*, John Wiley and Sons, New York.
5. Geertz, C.: 1973, *The Interpretations of Cultures*, Basic Books, New York.
6. Gelder, N. *et al.*: 1983, *Oxford Textbook of Psychiatry*, Oxford University Press, Oxford, England.
7. Kahana, R. J. and Bibring, G. L.: 1964, 'Personality Types in Medical Management', in N. E. Zinberg (ed.), *Psychiatry and Medical Practice in a General Hospital*, International Universities Press, New York.
8. Kimball, C. P.: 1981, *The Biopsychosocial Approach to the Patient*, Williams and Williams, Baltimore.
9. Knight, J. A.: 1991, 'Judging Competence: When the Psychiatrist Need, or Need Not, Be Involved', in this volume, pp. 3—28.
10. McCall, R. J.: 1975, *The Varieties of Abnormality*, Charles C. Thomas, Springfield, Illinois.
11. Millon, T. and Millon, R.: 1974, *Abnormal Behavior and Personality*, W. B. Saunders Company, Philadelphia.
12. Morreim, E. H.: 1991, 'Competence: At the Intersection of Law, Medicine and Philosophy', in this volume, pp. 93—125.

13. Pellegrino, E. D.: 1991, 'Informal Judgments of Incompetence: The Patient, the Family, and the Physician', in this volume, pp. 29—45.
14. Robertson, J. A.: 1991, 'The Geography of Competence', in this volume, pp. 127—148.
15. Schwab, J. T. and Brosin, H. W.: 1968, *Handbook of Psychiatric Consultation*, Appleton-Century-Crofts, New York.
16. Schwarcz, G. and Halaris, A.: 1984, 'Identifying and Managing Borderline Personality Patients', *American Family Physician* **29**, 203—208.

JUDGMENTS ABOUT PATIENT COMPETENCE: CULTURAL AND ECONOMIC ANTECEDENTS

The medico-legal-ethical concept of patient competence occasions more controversy than concensus. There is discernible agreement on just two points: (1) the criteria to be used in evaluating competence are extraordinarily elusive, and (2) patient competence plays a key role in medical decision-making.

Concern with patient competence is far from universal, however. The doctor/patient relationship in mainland China hospital *danwei* is highly asymmetrical and there is a preference for patients knowing as little as possible about their condition and treatment [24]. Those findings jibe with my experience (1980) visiting English-speaking Asian physicians in several Hong Kong hospitals: inquiries about patient competence and informed consent drew blanks. Non-recognition of the concepts appeared to be total, and explanation was not only futile but seemed to widen the gulf. Indeed, patient competence is irrelevant when patient education is disvalued and acquiescence in physician authority is assumed.

The contrast with American values is marked. This paper reviews why, in our constitutionally ordered society, competence matters so much while, at the same time, the search for a satisfying test of competence "is a search for the Holy Grail" ([40], p. 283). The paper offers, principally, a cultural perspective on competence. It is an anthropologist's approach to the contradictions, inconsistencies, and rationalizations that permeate the definition and application of competency criteria within one society, ours.

I try to explain why there is so great a risk that arguably competent persons will be involuntarily treated despite (1) legal theory which presumes that every adult is competent, and (2) numerous high court opinions asserting the patient's right not to be touched without his consent. In effect, the question addressed is why there is so much tolerance for risking the 'harm' of violating the constitutional rights of competent persons.

A speculative final section predicts that the society will become more willing to risk error that carries the opposite type of harm, i.e., *not treating* persons who refuse care but are possibly incompetent to refuse. I will examine cultural and economic trends that seem likely to shift risk preferences.

I. THE JEFFERSONIAN LEGACY

Evaluation of competency matters because, in legal theory, a competent adult patient may usually refuse any and all proposed medical or nursing interventions. Many case law expressions of Thomas Jefferson's guiding constitutional principles, liberty and individuality, reflect respect for privacy, including patients' rights not to be treated without consent ([10], [11], [14], [16], [17], [19], [23], [32], [33], [35], [38], [41], [42], [43], [53]). Conditions under which competent persons' rights in this domain have been curtailed in courts of law are extraordinarily circumscribed ([1], [3]).

The most frequently cited restrictions of competent persons' right to refuse appear in Jehovah's Witness cases. Common threads in this litigation are that the disputed procedure (blood transfusion) is considered both routine medical practice and life-saving, and that one or more troubling conditions have also been present. The latter include findings that the decision to refuse care was ambiguously made (e.g., a competent patient indicating that he welcomed court-ordered care), occurred in an emergency setting, or adversely affected the well-being of a minor ([20], [26], [34], [46], [50]). The existence of a minor is not controlling: in a 1972 case, a Witness was allowed to refuse transfusion when the court found that his two young sons had been well provided for so they would not suffer unduly if the patient, their father, died ([3], [32]).

Nothing in these cases suggests that the patient's 'best interest' can be grounds for overriding a sincere and competently given refusal of care. In effect, the Jehovah's Witness cases present extraordinary circumstances that serve to refine and fix the boundaries of rights of privacy and consent.

Mental illness is another circumstance where an arguably competent patient's right to refuse care appears limited. In the specific instance where the patient is *both* mentally ill *and* presents an immediate threat of harm to self or others, most state laws provide for emergency commitment to a mental hospital. However, the further issue of involuntary treatment while involuntarily hospitalized is not settled thereby in most states: competency hearings with periodic review provide the involuntary mental patient with due process protection ([29], [53]). In Utah, where the commitment criteria are themselves very strict and incompetence is essentially documented in the course of commitment proceedings, one case suggests that due process requirements are thereby satisfied and treatment can be administered in the face of patient refusal [29].

The specificity of these conditions again serves to affirm that the competent patient's right to refuse unwanted care is nearly absolute. This constitutional protection is bolstered by the presumption, elaborated in case law, that the patient is competent; that is, the burden of proof falls on those who allege that the patient is incompetent. Overcoming this presumption should be difficult, in theory, because the standard of proof required is "clear, cogent and convincing evidence" [21]. This standard is intermediate between the "preponderance of evidence" of civil law and the "proof beyond reasonable doubt" of criminal law.

Capacity to understand, rather than the understanding itself, is emphasized in another opinion:

... does the patient have sufficient mind to reasonably understand the condition, the nature and effect of the proposed treatment, attendant risks in pursuing the treatment, and not pursuing the treatment... ([42], p. 367).

The *rationality* of the patient's decision is explicitly not a test of competence. A lower court opinion, in the context of asserting the patient's right to be informed, states:

The interest of the court is that of having the patient informed of all the material facts from which he can make an intelligent choice, regardless of whether he in fact chooses rationally ([16], p. 1653).

Other decisions affirm that it is inadmissible to enter the substance of a decision, or outcome, as a criterion that determines the decision-maker's competence. A 1964 opinion by Judge Warren Burger, later to be Chief Justice of the United States Supreme Court, addresses this point. Burger writes:

Mr. Justice Brandeis, whose views have inspired much of the 'right to be let alone' philosophy said ...: "The makers of our Constitution ... sought to protect Americans in their beliefs, their thoughts, their emotions and their sensations. They conferred, as against the government, the right to be let alone ... the most comprehensive right and right most valued by civilized man" ([31], p. 438). Nothing in this utterance suggests that Justice Brandeis thought an individual possessed these rights only as to sensible beliefs, valid thoughts, reasonable emotions, or well-founded sensations. I suggest he intended to include a great many foolish, unreasonable and even absurb ideas which do not conform, such as refusing medical treatment even at great risk ([20], p. 978).

Thus, legal theory places the burden of clear and convincing proof on those who seek to overwhelm a patient's decision for himself. Moreover, the irrationality of that decision is explicitly excluded as a criterion to be used in evaluating the patient's competence to decide.

For those cases where there is an inevitable risk of error because the patient's mental status is marginal, this body of law might be seen as directing a preference for finding that the patient is competent. This would mean allowing an incompetent patient to refuse needed care rather than risk forcing treatment on a competent patient who has refused it. That is, the nation's highest courts appear to have come down on the side of safeguarding liberty.

Put more strongly, one should be surprised to find even a single case where extended litigation results in court-ordered nullification of an arguably competent patient's refusal of care.

That was the outcome, however, in one well-documented case. Here, an elderly woman who was described by all as coherent, articulate, and alert was not allowed to refuse bilateral below-the-knee amputation. In key testimony, a psychiatrist asserted that she was using "psychotic denial" to avoid confronting the seriousness of her medical condition. He further stated that, although she was incompetent to refuse, she would be competent if she consented to surgery ([2], [4]).

II. THE REALITY

With minimal sensitivity, one soon notices that it is routine to force unwanted care on arguably competent patients. One need not work near a hospital or court of law to learn that there is an overwhelming preference for risking the error of forcing care on a patient who actually is competent. In the usage found in this volume, 'beneficence' prevails over liberty. The risk of error established by due process in criminal law, allowing numerous guilty to escape punishment so that one innocent person is not wrongly deprived of liberty, does not obtain.

A patient's liberty is not so valued, apparently, as his life. The irony is the greater because so often that remaining life has miniscule potential for quality, particularly when one has wrenched away the liberty. 'Beneficence', also, is a misnomer: why, for example, should amputation (even if it carries a better chance for life) be preferred to conservative care? Amputation may be anathema to some patients. Moreover, it frequently leaves the patient with severe 'phantom' pain from the amputated limb, wreaks havoc with self-image, leaves the patient much more dependent on others, and is itself neither a benign procedure nor reversible.

Explanation is due on why the preference for risking harm to liberty, so contrary to constitutional intent and landmark case law, has gained currency.

III. CULTURAL PERSPECTIVE ON COMPETENCY CRITERIA

Judgments about patient competence are embedded in the culture. A corollary is that the criteria used for evaluating competence adjust in order to produce outcomes that accord with the society's value system and current requirements for order and economy. Since it inheres in these criteria to shift adaptively in response to cultural and economic realities, it follows that a search for a stable and satisfying test of competence is futile.

Liberty and the avoidance of medically defined harm often compete for priority in competency decisions. Liberty issues have constitutional implications, and yet decisions about competence reflect not only law, but also factors such as the reward structure of medicine, authority relations among various sectors of the society, and a culture's hierarchy of values.

Judgments about competence appear to be responsive to the national respect for technology and scientific expertise. The authority granted to physicians is one aspect of this larger theme and results in priorities defined by physicians being usually dominant in determinations of patient competence. The deference accorded physicians has also been a factor in retarding the development of due process procedures to protect patient rights.

The authority granted to physicians by the society at large can be easily demonstrated. For example, it is strikingly clear that physicians successfully assert influence in many areas for which they have no special training or expertise. Such domains include child rearing, nutrition, rehabilitation of criminals, city zoning, and education. In the public mind, physician opinion is equated with expert opinion, no matter that a particular physician has taken up an issue merely as a sideline or avocation.

The area of real physician expertise is, of course, medicine. In this realm, patients often act as though persuaded that physicians speak gospel (sometimes attributing certainty that a physician himself does not intend, and demanding reassurance that a physician cannot give.) Until recently, third party payors have been equally respectful of medical judgments.

Payors or legislators who resist medical priorities for patient care have been easily branded as inconsiderate of the needy or not compassionate. Patients who question the wisdom of a procedure fear more severe penalties. Realistically or not, patients perceive abandonment as an implicit threat that deters them from objecting too determinedly to a treatment plan. Those who brave dispute and abandonment may encounter the display of medical power that is the *raison d'etre* of this volume: a challenge to their competence.

This volume attests to physician authority in more ways than one. I am struck that several authors accept a self-serving medical definition of the debate: opposition of liberty and beneficence. The dictionary meaning of 'beneficence' is the state or quality of "doing or producing good, especially performing acts of kindness and charity" [51]. Thus, beneficence in our context attributes a motive, doing good, and at the same time describes avoiding physical harm befalling the patient, the harm being medically defined. It entails treating the patient for his own good with doctors deciding what the good is. What happened to the old word, paternalism?

Given that physicians' values have penetrated deeply into the culture, those values should be examined. Eminent psychiatrist Marc Hollender has addressed the exact issues debated here, writing:

... I favor thinking first of social responsibility rather than individual liberty. Certainly social responsibility, as a value, encompasses survival and compassion . . . ([25], p. 962).

It is noteworthy that Hollender's paper was invited for juxtaposition with one from Thomas Szasz, a noted libertarian psychiatrist. Articulate and a prolific writer, Szasz stands outside mainstream psychiatry and medicine precisely because of his assertion of patient responsibility and autonomy and his disparaging view of psychiatry as a tool of social control ([47], [48], [49]).

Hollender's allusion to "survival" flags another value that is at the core of twentieth century medical tradition. Life is everywhere a recognized good, and in the United States it is regarded more highly than in many societies. But even within a culture that highly values life, physicians stand out, I think, as extreme.

For example, research suggests that nurses and doctors differ on their views of death. Nurses admit to redeeming aspects of death, including cessation of pain and security in the sense of safety from the torturous uncertainties of disease. Physicians tend to condemn every aspect of death, emphasizing its coldness and aloneness. Moreover, many physicians view a patient's death as a personal and professional failure ([13], [27], [52]). They are less likely than nurses to identify comfort for the dying as a professional objective. In short, physicians appear to feel relatively more compelled to resist, with any means or cost, a patient's dying [12]. (Whether this pattern exemplifies vitalism is a question for theologians.)

Insofar as nurses are also professionally dedicated to the patient's good and, moreover, make vastly greater time commitments to the individual patient, nursing views may have greater validity. It might not be rash to conclude that treatment decisions left to physicians will be systematically flawed by insensitive over-valuing of life.

This speculation is supported by the record of families who have sued to stop aggressive treatment of relatives ([6], [12], [15], [18], [35], [38]) or who have felt driven to record their helplessness to stop inhumane care [45].

Added to physicians' commitment to prolonging life is a bias toward active intervention in the disease process. Within medicine, battles over homeopathic therapies and the more passive stance of helping the body to heal itself were fought and lost nearly a century ago [44]. The tradition of respecting and supporting the body's own healing powers is alive and well primarily through nursing.

The preferences inherent in modern medicine have been exaggerated by a variety of incentives. Physicians themselves acknowledge the threat of malpractice litigation as an inducement to overdiagnose and overtreat. In addition, both research and medical teaching objectives are best served by active interventions and prolonging every life. Furthermore, tests and procedures are financially more rewarding than passive approaches to care.

Central to our topic, psychiatrists are frequently cast as arbiters of patient competence and, as a consequence, are subject to additional incentives generated from within medicine. Requests for evaluation of a patient usually come from other medical specialists, and the patient population referred is *not* representative of all medico-surgical patients whose mental status is marginal. On the contrary, psychiatrists are called primarily to evaluate patients who do not consent to a recommended procedure.

A psychiatrist's report that a patient is competent usually halts thought of legal proceedings to appoint a guardian and get substituted consent. Thus, psychiatrists are gate-keepers who can frustrate other specialists in their drive to treat. The unique incentives that psychiatrists face are, as a result, mostly negative and exclusionary.

Frank S. Ranuska comments,

> In my experience as a psychiatric consultant, no cases have caused as much alienation of medical staff and the psychiatric service as finding a patient competent to refuse treatment. After such evaluations, requests for psychiatric consultation have dramatically decreased in number, with some medical staff viewing psychiatry as oppositional to good medical care ... (unpublished letter to the editor of the *American Journal of Psychiatry*, quoted with the writer's permission) [37].

Ranuska's point corresponds to my own observation of psychiatrists' beliefs about cause and effect: if psychiatric activities in support of a patient's independence infringe too much on other specialists' turf, referral sources will dry up. Loss of referrals weakens the psychiatry consultation-liaison service in a hospital setting; in private practice, it has a direct impact on income.

Psychiatrists generally share mainstream medical values so here, as in other specialities, predispositions and the incentive structure converge. Not surprisingly, the psychiatric literature on competence is tilted toward paternalism.

As Beauchamp documents in this volume, the 'consequentialist' approach is widespread in psychiatry [8]. Now, the essence of the consequentialist principle is to risk a harm to liberty when the threatened physical harm (from not overwhelming the patient's refusal) is great. Insofar as draconian treatments are often medically indicated when life is at stake, consequentialism assaults the patient's liberty over the very procedure the patient may most want to refuse. Autonomy is an empty concept if highly consequential decisions are precisely those that trigger use of more rigorous competency criteria, increasing the chance that the patient will not be allowed to choose for himself.

In addition, there is a Catch-22 flavor to psychiatric reasoning on competence. As noted above, a psychiatrist testified in a court of law that a patient would be competent if she consented [to amputation], but was incompetent so long as she refused ([2], [4]). Such a conclusion is rationalized in the psychiatric literature. Applebaum and Roth write, for example,

If a treatment or procedure for which consent is being sought is sufficiently provocative of anxiety and fear, the patient may be forced to revert to more primitive, even psychotic, levels of defense for coping with it. Thus, the recommendation for the procedure itself may force the patient into an apparently incompetent state ... ([5], p. 1463).

Jargon aside, this means that a patient who is scared silly of a treatment may be denied the right to refuse that treatment because his fright renders him 'incompetent'. But fear signifies the patient's antipathy for a particular course of action. What becomes of the Burger-Brandeis doctrine if fear, even fear resulting in decision that appears irrational to others, itself becomes a basis for overwhelming the presumption of competence?

A further problem with psychiatric evaluation of competency is that data is obtained in an inherently coercive context. Patients who will not cooperate with a psychiatric interview are suspect on these grounds alone. Moreover, psychiatrists invariably represent themselves as coming to 'help', and often neglect to clarify that the data they gather might be used in an adversarial process. Judge David Bazelon remarks of this failure to warn, "For the psychiatrist to encourage a free flow of communication from the patient as though the traditional doctor-patient relationship existed is nothing short of fraud" ([7], p. 914).

Although a hospitalized patient is inevitably disadvantaged by institutional structures, the balance can still be rectified when conflict over competence moves to a legal arena. However, to the extent that medical values are shared by the community at large, medical authority goes unchallenged in the courtroom. Particularly when patient competence is at issue, deference to physicians is patent.

Some deferential patterns seem counter-productive because they subvert fact-finding, which is a primary court function. Fact-finding is impaired when courts too readily accept medical assertions that a crisis is at hand, when physician testimony consists of conclusions instead of supporting facts, and when 'expert' opinion is seen as more reliable and intrinsically superior to a patient's intuitions about his own body.

Courts are not designed to respond to crisis, but judges take seriously a surgeon's assertion that the patient must be operated on immediately. In the *Northern* case, a Chancery Court judge held a hearing before regular court hours, accepted a stay while a psychiatrist fought through a Nashville, Tennessee, blizzard to evaluate the patient, himself drove through unplowed, icy streets to collect said evaluation, and reconvened court within 24 hours. This judge later said that he felt as though the surgeons were waiting by the telephone, scalpels in hand, for his decision. On appeal in the same case, three judges held a Saturday morning bedside hearing. The *guardian ad litem* was appointed by the Court minutes before the first hearing, which was four days before the Court of Appeals convened at the bedside ([2], [4]).

The 1984 case of *Bertha Harris* has similar elements. The issue was Harris' competence to refuse amputation of her (remaining) leg. The *guardian ad litem* was an attorney relatively new to practice and was appointed five minutes before the court was to convene ([9], [30]).

A difference in *Harris* was that, serendipitously, the *guardian ad litem*, Barbara Mishkin, had been on the President's Commission for the Study of Ethical Problems in Medicine and Biomedical and Behavioral Research. Knowledge of the relevant issues and law enabled Mishkin to be an effective advocate. Her brief alludes to the air of crisis as one element that robs patients of their constitutional rights:

Even within the jurisdiction of this Court, the case of Bertha Harris is not an isolated event. We are aware of at least 40 cases (in addition to the one at bar), adjudicated during a period of 18 months, in which the court has been petitioned to declare a patient incompetent and to authorize medical interventions.

Just as in the case of Bertha Harris, petitions typically move rapidly through the Superior Court. For example, in the case of *Earl Lewis*, the right arm was removed on an emergency basis on May 2, 1983 (without court authorization). On May 27 (the Friday of Memorial Day Weekend), the hospital petitioned Superior Court for authority to remove Mr. Lewis' right leg. A hearing was held by Judge Fauntleroy on May 31, which was the Tuesday following Memorial Day and the same day that the court appointed a guardian ad litem to represent the patient. Mr. Lewis' right leg was amputated the next day (June 1). On June 3, the hospital again sought authorization for surgery on an emergency basis, this time to remove Mr. Lewis' left leg. A new guardian ad litem had to be appointed because the one who had represented the patient at the first proceeding was unavailable. The second guardian's report to the court was submitted on June 6, the day that the third amputation was scheduled to take place.

As is evident ... court-appointed counsel are at a serious disadvantage in providing effective representation of their clients, even in the Superior Court. That the cases generally evade appellate review is an understatement of enormous proportions ... the constitutional rights of scores of patients in the District of Columbia are in serious jeopardy ([30], p. 8—9).

And what of patients in 50 states?

The air of crisis is just one condition that diverts the court from its fact-finding mission. Another is that physician testimony frequently takes the form of conclusions bare of supporting facts. *Every diagnosis is a conclusion.* Depression, anxiety reaction, and psychotic denial are all conclusions, and they often are used authoritatively as the basis for declaring a patient incompetent. The psychiatrist in *Northern* asserted that the patient was using "psychotic denial", but made scant effort to report her words, in context, that had led him to this determination.

Bazelon is broadly critical of psychiatric use of conclusory labels. He states that the practice thwarts cross-examination, which "is the law's principal mechanism of sifting through competing facts and values presented in the court" ([7], p. 916). Bazelon stops just short of imputing intent: "Different jargon and different labels are used with the same effect: concealment of the difficult issues of the state's power to deprive individuals of liberty" ([7], p. 915).

Judges do not usually accept conclusions in lieu of facts. That they routinely accept this type of testimony in competency proceedings speaks both to the air of crisis in these cases and to the deferential treatment accorded physicians.

A conclusory type of testimony, unsubstantiated by facts, is no better than opinion and, on its face, has no greater validity than a patient's opinion. Nevertheless, patients' opinions are commonly called intuitions and are discounted.

Intuitions about one's own body easily merge with a will to get well, or contrarily, with giving up. These decisions that a patient makes about himself are known to change the course of disease and even the ultimate prognosis [22].

Hope and a will to recover are facts in that they tell something about the patient. As prognostics, they may have as much validity as expert conclusions that are unsubstantiated by facts. Nevertheless, a patient's intuitions and hopes are sometimes treated as one more evidence of irrational thinking and, *ipso facto*, incompetence ([2], [4]).

In short, patients who refuse to consent to treatment are likely to find that medical authority extends beyond the hospital setting. Medical values have penetrated the culture into the courtroom. The disease model of medicine with its requirement for active intervention in order to prolong life has largely gone unchallenged. A view that life is beyond price has evolved into 'treat at any cost'. Since the financial cost has typically been born by third-party payors, and the physical and emotional costs are indubitably the patient's and family's, there have been scant brakes on excess from within the medical profession.

IV. TRAILS TO FOLLOW

Culture lags behind changed circumstance, as a rule. Nevertheless, cultures cannot indefinitely sustain disjunctions between closely related and highly visible themes. Adaptation is slow; but it is, nevertheless, inexorable. Ethical and legal concerns with patient autonomy may be the cutting edge of an adaptive process that is under way.

Medical authority will be challenged without doubt if it conflicts with other values and needs in our society. If this eventuates, the 'outcomes test' of patient competence will shift to subserve the new conditions. My concluding remarks suggest sources of conflict that will change medicine from within, and/or will materially restrict its present eminence.

Although it seems preferred, currently, to risk a harm to liberty rather than medically identified harm, I expect change. Three developments will inspire different balancing of these risks.

There is, first, the growing disenchantment with some aspects of medical technology and a corresponding willingness to question whether prolonging life under certain circumstances benefits the patient. Popular interest in Living Wills and state legislation that gives them legal force is but one manifestation of consumer concern, a dread lest patients and families lose control of decisions about care. Life-prolonging technology applied in the modern hospital is particularly feared because it is perceived to be indiscriminately used without regard for quality of life and other values.

The significance from the anthropologist's perspective is that the fear and knowledge of abuse are not isolated; they are broad enough to trigger legislation! This questioning mode has infused the culture and seems unlikely to reverse so long as the capacity to restore a semblance of health is so far outpaced by the means to prolong 'life'. I think that this distrust of some medical practice will generalize, creating a predisposition to see the reasonableness of a patient's refusal of care.

Second, the new emphasis on individual responsibility for maintaining health will probably enlarge the support base for patient retention of control during illness episodes. Persons who daily renounce gratifications in order to promote their own health are not likely to be tame patients. Individuals who expect an advisory rather than authoritarian stance in their own physicians should have quicker empathy for others' protestations against unwanted care. This again is a cultural process.

Third, limited funding is constraining health care in ways that will be perceived as shortage, bringing new perspectives to bear on waste of all sorts. Dramatically expensive procedures are not the only wanted interventions unavailable to many. In Tennessee, Medicaid pays just 21 days of hospital care per annum for a medically indigent adult, and municipal hospitals primarily serving indigent populations remain in constant financial jeopardy. Nationally, Medicare's Diagnostic Related Group (DRG) method of calculating hospital reimbursement has created pressure to shorten length of stay. Similarly, employers are negotiating 'preferred provider' contracts that reduce hospital charges and also link employee health insurance coverage to compliance with various cost-containment strategies.

Physicians nationwide also feel the resource pinch. For example, many refuse to 'accept assignment' of Medicare fees, preferring to collect directly from the patient who then is reimbursed the 'covered' portion. The government countered (Spring, 1984) with punitive disincentives for *not* accepting assignment. Medicare patients fill about 30 percent of hospital beds, so physicians are under pressure to continue caring for them at the same time that they may feel inadequately compensated for their own services. Medicare beneficiaries are themselves faced with higher premiums for Part B of Medicare, which covers physician services. These or other forms of health care rationing will be a spur to re-examine the merits of forcing care on persons who would rather go without.

In summary, disenchantment with medical technology, beliefs that persons are individually responsible for maintaining health, and perception of resource scarcity may sharpen skepticism of medically defined harm and benefit. Those who are sick, who pay for care of the sick, who preside over determinations of whether a patient is competent to control his own care, or who decide for patients who are incompetent will all approach health care decisions with somewhat more independence. I do not doubt there will be change; I question only its extent and the immediacy of its impact on evaluation of patient competence.

Reiterating the general point that judgments about competence reduce to what type of harm the society is more willing to risk, I will restate my prediction. I expect increasing reluctance to let physicians have their way with patients by overwhelming a patient's refusal of a procedure or treatment plan. Given the values and mechanisms of our society, one way to allow a patient to prevail is to find that he/she is competent. Those elusive competency criteria will still be applied with an eye to outcome, but the preferred outcome will be to support patient refusal of care.

Perhaps you are sorry you asked an anthropologist.

Vanderbilt University School of Medicine
Nashville, Tennessee

BIBLIOGRAPHY

1. Abernethy, V. (ed.): 1980, *Frontiers in Medical Ethics: Applications in a Medical Setting*, Ballinger Publishing Company, Cambridge.
2. Abernethy, V.: 1984, 'Compassion, Control and Decisions about Competency', *American Journal of Psychiatry* **141**(1), 53—57.
3. Abernethy, V. and Lundin, K.: 1980, 'Patient Competence: Conditions for Giving or Withholding Consent', in V. Abernethy (ed.), *Frontiers in Medical Ethics*, Ballinger Publishing Company, Cambridge, pp. 79—98.
4. Abernethy, V. and Lundin, K.: 1981, *The Mary Northern Case: Public Compassion and Private Rights*, Case 9-481-690 and Teaching Note 5-482-635, Harvard Business School Case Services, Harvard University, Boston.
5. Applebaum, P. S. and Roth, L. H.: 1981, 'Clinical Issues in the Assessment of Competency', *American Journal of Psychiatry* **138**, 1462—1467.
6. Bartling v. Superior Court: 1984, 163 Cal. App. 3rd 186.
7. Bazelon, D. L.: 1978, 'The Psychiatrist in Court', in J. P. Brady and H. K. H. Brodie (eds.), *Controvery in Psychiatry*, W. B. Saunders Company, Philadelphia.
8. Beauchamp, T. L.: 1991, 'Competence', in this volume, pp. 49—77.
9. In the Matter of Bertha Harris: Washington, D.C. 1984, 477 A.2d 724.
10. Bouvia v. Superior Court: 1986, 179 Cal. App. 3rd 1127.
11. In re Estate of Brooks: Ill. 1965, 205 N.E. 2d, 442.
12. Brophy v. New England Hospital: Mass., 1986, 497 N.E. 2nd 626.
13. Campbell, T. W., Abernethy, V. and Waterhouse, G. J.: 1983—84, 'Do Death Attitudes of Nurses and Physicians Differ?', *Omega* **14**(1), 43—49.

14. Canterbury v. Spence: 1972, 464 F.2d 772, 781 (D.C. Cir.), *cert. denied*, 409 U.S. 1064.
15. In the Matter of Claire Conroy: New Jersey 1985, 486 A.2d 1209.
16. Cooper v. Roberts: 1971, 286 A.2d 647, 650.
17. Department of Human Services v. Northern: 1978, 563 S.W.2d 167 (Tn. app. 1978) *cert. denied*, U.S.
18. Dockery v. Dockery, No. 51439 (Ch. Ct., Hamilton City., Tn., filed February 11, 1977).
19. Erickson v. Dilgard: 1962, 252 N.Y.S.2d, 705.
20. Application of President and Directors of Georgetown College: 1964, 331 F.2d 1000, (D.C. Cir. 1964), cert. denied, 377 U.S., 978.
21. Grannum v. Berard: Wash. 1967, 422 P.2d, 814.
22. Hackett, T. P. and Cassem, N. H.: 1974, 'Development of a Quantitative Rating Scale to Assess Denial', *Journal of Psychosomatic Research* **18**, 93—100.
23. Harper, F. and James, F.: Supp. 1968, *The Law of Torts*, vol. **2**, 61 n. 15.
24. Henderson, G. E. and Cohen, M. S.: 1984, *The Chinese Hospital: A Socialist Work Unit*, Yale University Press, New Haven.
25. Hollender, M. H.: 1978, 'Should Psychiatric Patients Ever Be Hospitalized Involuntarily? Under Some Circumstances — Yes', in J. P. Brady and H. K. H. Brodie (eds.), *Controversy in Psychiatry*, W. B. Saunders Company, Philadelphia.
26. John F. Kennedy Memorial Hospital v. Heston: N.J. 1971, 279 A.2d 670.
27. McCue, J. D.: 1982, 'The Effects of Stress on Physicians and Their Medical Practice', *The New England Journal of Medicine* **306**(8), 458—463.
28. Meisel, A., Roth, L. H. and Lidz, C. W.: 1977, 'Toward a Model of the Legal Doctrine of Informed Consent', *American Journal of Psychiatry* **234**, 285—289.
29. Mills, M. J., Yesavage, J. A. and Gutheil, T. G.: 1983, 'Continuing Case Law Development in the Right to Refuse Treatment', *American Journal of Psychiatry* **140**(6), 715—719.
30. Mishkin, B.: 1984, *Opposition of Bertha Harris to Appellee's Motion to Dismiss Appeal*, the District of Columbia Court of Appeals No. 84-671.
31. Olmstead v. United States: 1928, 277 U.S., 438, 479.
32. In the Matter of Osborne: D.C. Cir. 1972, 294 A.2d, 372.
33. Perlmutter v. Florida Medical Center et al., Cir. Ct. Broward County, Fla., July 13, 1978, No. 78-9747.
34. Powell v. Columbia Presbyterian Medical Center: 1965, 267 N.Y.S.2d, 450—451.
35. Matter of Quinlan: N.J. 1976, 355 A.2d 647.
36. Raleigh-Fitkin Hospital v. Anderson: 1964, 42 N.J. 421, 201 A.2nd 537.
37. Ranuska, F. S.: 1984, unpublished Letter, January 13, Oakland, California.
38. In re Requena, Morris Co.: New Jersey, 1986, P-326-86E.
39. Roth, L. H. *et al.*: 1982, 'The Dilemma of Denial in the Assessment of Competency to Refuse Treatment', *American Journal of Psychiatry* **139**(7), 910—913.
40. Roth, L., Meisel, A. and Lidz, C. W.: 1977, 'Tests of Competency to Consent to Treatment', *American Journal of Psychiatry* **134**, 278—284.
41. Sard v. Hardy: MD 1977, 379 A.2d 1014, 1020.
42. In the Matter of Schiller: N.J. Super. 1977, 372 A.2d, 360, 367.

43. Schloendorff v. Society of New York Hospital: N.Y. 1914, 105 N.E. 92.
44. Starr, P.: 1982, *The Social Transformation of American Medicine*, Basic Books, New York.
45. Stinson, P. and Stinson, R.: 1983, *The Long Dying of Baby Andrew*, Little, Brown and Company, Boston.
46. State v. Pemberton: N.J. 1964, 201 A. 2d 537, *cert. denied*, 377 U.S.
47. Szasz, T. S.: 1961, *The Myth of Mental Illness*, Hoeber-Harper, New York.
48. Szasz, T. S.: 1963, *Law, Liberty, and Psychiatry*, MacMillan, Inc., New York.
49. Szasz, T. S.: 1969, 'Alcoholism: A Socio-Ethical Perspective', in *Transactions, Journal of the Department of Psychiatry*, Marquette School of Medicine **1**(1), 7—12.
50. United States v. George: D. Conn. 1965, 239 F. Supp. 752—753.
51. *Webster's New Collegiate Dictionary*: 1975, G. and C. Merriam Company, Springfield.
52. White, L. P.: 1977, 'Death and the Physician: Mortuis Vivos Docent', in H. Feifel (ed.), *New Meanings of Death*, McGraw-Hill Book Company, New York.
53. In re Yetter: C. P. Northhampton County, 1973, 62 Pa.D.&C. 2d, 623.

PATIENT FREEDOM AND COMPETENCE IN HEALTH CARE

If the new ethos of medical ethics is clear about and unreservedly committed to anything, it is to the proposition that competent patients must be allowed to retain their right to manage their own affairs within health care, and the legal and moral doctrine of informed consent simply seeks to protect and enshrine this right. Given this almost absolute provision (certain exceptions such as the emergency treatment of suicides have been generally seen as permissible within the rule), one might well wonder what confusion (or incipient paternalism) has generated our volume's focus, 'when are competent patients incompetent?'.

The sense and legitimacy of our issue is easily gained, however. Noting that most patients seem to satisfy the low-threshold criteria for *general competence* (in effect, they are generally oriented, and present with no glaring suggestions of incompetency, such as psychosis or severe retardation), our issue arises simply by noting certain other clinical realities that cast doubt on the actual ability of a significant number of 'competent' patients to perform the task the new ethos has set them, *viz.*, to make their own decisions about their medical care. First, there are clearly certain factors, often coincident with illness, that may undermine or diminish (though not necessarily obliterate) a given patient's cognitive and volitional abilities, e.g., fear, stress, pain, the effects of the treatment (drugs), and the effects of the pathology (lessened oxygenation of the brain secondary to pulmonary deficits). Second, whether the previous sorts of factors are presented, certain patients may also be significantly flawed in their understanding of their situations and prospects, as suggested by the presence of denial, false beliefs, or by recent studies showing poor recall of seemingly adequate informed consent by a significant number of patients who were tested post-operatively [8]. Third, even if the first two sorts of factors are not apparent, all the rhetoric extolling patient autonomy will probably not afford us complete comfort when faced with a patient who rejects a lifesaving treatment, or chooses what we believe is a less than optimal (or 'unduly' risky) treatment option. Thus, given the presence of such troublesome 'clinical realities', our issue arises in terms of two quite different senses of the word 'competence': (1) competence as a general moral and legal status of freedom and autonomy that most adults enjoy, and (2) competence as the ability to perform a task. Most patients seem to merit the former general status, but we should be concerned that a significant number of them, because of the three sorts of clinical realities just noted, may not, in fact, be able to perform the task at hand adequately, i.e., to make a specific medical decision.

Now, by way of anticipation, I submit that this volume's focus on the question: 'when is a competent patient incompetent?', must be seen not only as raising the previous concerns but also as intentionally (and provocatively) begging a very crucial question. The question begged is: should the clinical realities just noted be allowed to drive us to operationalize certain tactics of monitoring and testing patients' actual decision-making that such patients, since they are, by hypothesis, still of the class of the general competence, would *not* be subject to in most other areas of human endeavor? In effect, should we hold certain patients to a more stringent 'to the task' test of competence even while they retain general competence? My own view here will be negative, and in two senses: (1) that such heightened monitoring and testing fundamentally contradict the basic principles and insights of the new ethos which I take my colleagues in this volume to be proponents of, and (2) that the strength of the new ethos is sufficient to deny the appropriateness of any restructured and rival medical ethics that proceeds in such a vein.

THE NEW ETHOS OF MEDICINE

As a proponent of the new ethos, I have myself asked (as if the answer were clear), whether there is "any good reason why clearly competent people somehow should lose their right to maintain control over their lives simply because they become sick and come under a physician's care?" ([10], p. 254). The answer seems to be clearly negative. The right of patients to make their own decisions in medicine is a *regulative principle* (or "first order principle" as Robertson puts it ([7], p. 127), and picking out incompetent patients or those patients of substantially diminished competence is merely a subsidiary 'tidying up' operation, having no implications for the vast majority of patients.

The grounds for enshrining this principle seem quite straightforward and unimpeachable. Perhaps the strongest argument is the one that proceeds by simply pointing out that all competent people are free to manage their own affairs in all other areas of human endeavor, so the burden of proof is on those who would make an exception to this within the context of health care. The proponent of the new ethos thus immediately assumes the high ground, and the detractor is burdened with the task of assaulting a well-entrenched moral, legal, and social principle.

Moreover, the assaults have generally seemed to falter. The argument that medical decision-making is much more complex than in other areas is probably false if one considers what is implicit in, for example, getting married, choosing a career, or building one's estate. Equally, this argument ignores the point that we all make many decisions on the basis of understandings that would pale alongside of those of an expert in the area, e.g., purchasing stocks or a home. But this is hardly regrettable; rather, our right to make such decisions, however unsophisticated, is a necessary condition for our very freedom. To require expertise in any area would deprive most of us of our freedom in that area. Another argument, *viz.*, that medical decisions are much more serious and potentially disastrous, seems equally ineffective. We do not stop the mountain climber, the stuntman, the smoker, or the alcoholic, though we know that they clearly are running risks that are much more serious than in most health care decisions. Equally, it seems that part of being free is the freedom to make choices that appear foolish, stupid, or tragic to others — better to run the risks than allow Big Brother a veto. Further, I know of no research documenting the proposition that most, many, or even a substantial number of patients choose poorly or disastrously (in their own eyes *or* to others), however ignorant, confused or informed they are. Finally, whatever emphasis we may put on the spectre of patients with diminished competence, surely the general thrust of the preceding remains unimpeached. At least nine out of every ten patients are not even sick enough to be hospitalized, and whatever the 'competence-diminishing' factors they may be suffering from, for example, neuroses, compulsions, or anxieties, these factors are usually the same ones they carry with them through the rest of their day — surely we do not wish to impede their management of their affairs in general.

So: the new ethos triumphs. Even our present issue seems to present, at most, the need for a 'tidying-up' operation, the first option we should consider.

A. *Tidying-Up the New Ethos*

If we pause to review the writings of my colleagues in this volume, however, we may get an uneasy feeling. All are, I gather, proponents of the new ethos, at least in its general thrust, and seem, at first glance, to be merely disagreeing over specific tactics and criteria by which a follower of the new ethos can respond to patients who may present with one or more of our three troublesome 'clinical realities', i.e., factors that may produce a diminished competence, evidence of a flawed understanding of the situation, or questionable choices. But my colleagues' disagreements are, in fact, more basic and substantial than one might have anticipated they would be.

Both Drs. Knight and Pellegrino seem to see the issue of whether a patient is competent to decide as a frequent issue for all physicians ([2], p. 3; [5], p. 29), and Beauchamp's arguments would seem to make it so ([1], p. 55ff). Morreim, in contrast, would clearly reject any such pervasive and stringent 'gate-keeping' functions by the general practitioner because of the quite inconclusive implications of factors that diminish competence even when they are found ([3], pp. 97—98, 105—106). Fear, pain, or stress may diminish competence, or they may not — who is to say? As Pellegrino notes, some patients are still able to focus clearly and perform the decision-making task adequately ([5], p. 32). Second, regarding the significance of outcomes and risk-benefit ratios, Knight, Beauchamp, and Robertson recommend that these factors be integrated into competency testing directly, i.e., for lessening or heightening the requirements a patient has to meet to be deemed competent to make a specific medical choice ([1], p. 66, [2], p. 12, [7], p. 140, in brief, to the degree *we* see the patient's choice as 'favorable', the lower the test of competence; the more 'unfavorable', the more stringent the test). Pellegrino and Morreim, for their part, clearly reject any such reference and variability ([3], pp. 99—100, [5], p. 35). Finally, there seems to be a great deal of disagreement as to whether the competency at issue here is general competence, or the situational sort of competence to perform the task at hand. Morreim's essay is designed to push for the former, while Pellegrino and Knight clearly hold to the latter ([2], p. 3, [5], p. 30), as, by implication from their stand on the relevance of outcomes, do Robertson and Beauchamp. Pincoffs, for his part, would seem to be forced into the latter camp also on the basis of his remarks on "encumbrances" to patient decision-making ([6], p. 85), but I suspect (on the basis of his other remarks) that he would be unhappy with such a result.

My first point here is that such issues and disagreements bespeak much more than a 'tidying-up' operation. If my prior rendition of the highground stance of the new ethos is accurate, then such discussions are, in fact, outside its pale and represent options that it cannot countenance. For one thing, a movement to a 'to the task' orientation for competence judgments conflicts with the fundamental thrust of the new ethos that the right to manage one's affairs with minimal impedance and overseeing is global. Such freedom does not somehow embrace all other areas of human endeavor and stop short at medicine as if this were some essentially new sort of ball-game. In effect, general competence is a sufficient test, however low-threshold; to require more stringent tests is to depart from a precious norm without providing any good reason for doing so. Similarly, reference to outcomes and risk-benefit ratios in determining the looseness or stringency of tests of competence flies in the face of the point that the seriousness of outcomes, or potential for disaster, is generally not given such status elsewhere. In sum, I submit, the use of the 'to the task' sense of competence, and the reference to outcomes, contradicts the fundamental grounds of the new ethos. The new ethos, in effect, rests on the broader social meanings of competence and freedom, within which more stringent tests, or specific reference and evaluation of outcomes, is all but unknown. And, again, it seems that we must demand this to be so as a necessary condition for our freedom, both as patients and, to the extent we are concerned with slippery-slope arguments, in general.

My argument so far is thus: if we see the new ethos of medicine as a logical outgrowth and extension of what it means to be a member of a free society, then the attempt to monitor and test our three clinical realities (i.e., diminished competence, flawed understanding, and questionable choices) is not permissible. In effect, it is not just that the threshold of the right to manage one's own affairs is minimal, but that the right is so primary and important that we should not even allow for much in the way of overseeing its exercise, nor in evaluating the level of competence, understanding or 'appropriateness' behind any *particular* decision. In effect, the argument here echos Robert Nozick's remarks on freedom as a "side constraint" in ethics — it is not just another value to be factored into an ethical analysis, it is a primary regulative principle which must, in most situations, simply be respected. ([4], pp. 26—33).

But there is another option here that I sense my colleagues to be flirting with at least implicitly. That is: why not restructure the thrust of the new ethos in a major way, *viz.*, that there is, in fact, something special about being ill and being faced with medical decisions that makes health care *different in kind* from other areas of human endeavor? In sum, rather than see the new ethos as just the logical extension of our general social freedom, might we not instead restructure it to be a consistently argued exception to the rule?

B. *Restructuring the New Ethos*

This new option is what I take my colleagues in this volume (with the exception of Morreim clearly, and Pincoffs probably) to be exercising in one form or another. Further, I submit that they are essentially appealing to our three troublesome 'clinical realities' toward the establishment of such an 'exception to the rule' claim for health care: (1) that diminished competence is too pervasive and substantial in health care to ignore as we do its analogy in other areas (e.g., pick any of your neurotic acquaintances and review the nature and results of their recent life-decisions and behavior); (2) that ignorance, delusion, or the false comprehension of facts is much more rife in health care; and (3) that the outcomes in medicine are, in fact, often much more 'poorly', disastrously, inauthentically chosen by patients than elsewhere.

I submit, first, that I know of no studies that support any of *these* claims. Second, I strongly suspect that they are all false: (1) most patients are not, for example, hypoxic, terrified, or drug-intoxicated, and for every patient who *is* burdened by a substantial competence-diminishing factor, I suspect we could produce a dozen people from the society whose neuroses or compulsions are in full-swing, but whose activities are not being overseen or curtailed. (2) How much knowledge does the average car buyer bring to that task or, given the divorce statistics, how much mutual understanding is present at the usual wedding? As to (3), the lack of any supporting study is enough to dispense with this.

It will be objected here that the restructuring needed only relates to a *select group of patients* who are, in fact, diminished in their competence, and/or flawed in their understanding of the situation at hand, and/or making disastrous decisions. Operationally, I gather the tactic here would basically be that the clinician would check for any evidence which would suggest the *possibility* of an inadequate competence 'to the task'. Finding any such factors, a more formal evaluation would be 'triggered', perhaps utilizing a psychiatrist, toward a judgment of whether 'to the task' competence was 'adequately' present. My response is that to pick these people out we would have to subject a much larger number of patients to a scrutiny that is disrespectful and threatening of their freedom. We may talk here of 'triggering' such an inquiry selectively, e.g., if the patient is overly fearful, but to subject a fearful patient to such scrutiny, and to leave all those neurotics free to act out simply because they are not sick, clearly seems to be a gross injustice.

My concern here, in other words, is that aside from the problem of generating sufficiently precise operational criteria and cut-off points for our troublesome 'clinical realities', such policies are likely to result in a much more pervasive scrutiny of many patients. The poor recall studies of informed consent [8] suggest that many patients would have some flaws in their understanding and many of the factors that diminish competence are often present to some degree in all patients. Similarly, since many illnesses have multiple therapeutic options, any patient who chooses other than what the physician consideres the optimal treatment might well be under suspicion of incompetence. Thus, heightened scrutiny and testing would be triggered in many cases where I suspect we would want to say that even 'to the task' competence is retained. But such pervasive triggering raises serious questions regarding invasion of privacy, the abnormal impedance of freedom, and unjustly risks the generation of false negatives regarding patients who would retain their freedom elsewhere.

I respond, more generally, here in another vein. It would seem that the completely undocumented claim that a certain number of patients are choosing disastrously/inauthentically/poorly is an absolutely necessary part of this whole attempt at restructuring. Should we not simply say that freedom remains such an important value that we must demand proof of great and frequent disasters to even consider changing the rules in this one area? Let us grant that diminished competence and/or flawed understandings *are* much more frequent and substantial in health care than elsewhere. It would seem that Morreim and Pellegrino are correct in advocating a greater readiness to provide a *therapeutic* and *educational* response to such factors, if there is time, but this hardly justifies switching to a pervasive system of monitoring and testing competence, or of placing individual self-management in jeopardy. We should also want to have good reason to hold that individuals about to be determined to be incompetent are about to make a 'bad' decision.

My suggestion here thus parallels the usual formulation regarding involuntary commitment for mental illness, *viz.*, that, beyond documenting the presence of mental illness, significant 'danger to self or others' must also be established.

But what constitutes a bad decision in medicine? Here I am surprised and somewhat disheartened to see my colleagues write as if medical decisions can be evaluated in some straightforward, objective manner. Following the work of Roth and his colleagues, Knight says, without blinking:

> [I]f, for a serious illness, the risk in treatment is minimal and benefit is great, the patient who refuses treatment is likely to be declared incompetent. If risk is great and benefit is minimal, the patient who refuses treatment is likely to be declared competent ([2], p. 13).

This, of course, is not the prescription of Roth *et al.* Rather, they advocate more stringent testing of the competence to consent (to the specific task) to the degree that a *refused* treatment has a *favorable* risk/benefit ratio *or* a *requested* treatment has an *unfavorable* risk/benefit ratio ([9], pp. 283—284).

But whence, pray tell, cometh these judgments about how favorable or unfavorable a given risk/benefit ratio is? Roth *et al.*, and some of my colleagues here, seem to treat such a judgment as an objective one, easily generated and unproblematic. Here again, I submit, we find a tactic that is in basic contradiction to the new ethos. In fact, I had always thought that a particularly telling *subsidiary* argument for the new ethos was the *fact* that *all* medical judgments involve *value* judgments. We have only to look at the recent work in clinical decision analysis to see that *any* medical option must be evaluated by an integrated consideration of the *probability* times the *value* of the risks and benefits involved ([11], pp. 168—227). Thus, given our sense of the subjectivity of values, and the pluralistic nature of our society, it is the patient at risk that should be quantifying these values. In effect, even if we were *not* a free society, patient evaluation of risks and benefits would still be essential, and the judgments of a physician or any other proxy regrettable guess-work and questionable mind-reading at best. Who better to evaluate a risk than the person who has to run it? Who better to evaluate a possible benefit, *per se* and in terms of the risks involved in pursuing it, then the one who is to receive the benefit (or perhaps only the risk since the benefit may not occur). It is almost as if Roth *et al.* were willing to countenance autonomy and freedom in medicine only to the extent that they agree with the decisions generated. If the patient departs from the evaluation of the 'experts', however, then the patient had better be able to negotiate some pretty high hurdles (the higher the 'unfavorability' to the 'expert', the higher the hurdle). This hardly involves a respect for others' freedom and autonomy. It is, as a matter of fact, strong paternalism of the most extreme variety; it not only presumes to be able to know clearly what is best for a patient specifically when a patient is disagreeing, but explicitly advocates the right to overrule that disagreement even though, by hypothesis, the patient is competent in the only sense I know of when we are talking about freedom and the right to retain control of one's life.

Finally, I find myself wondering, as Robertson does in his last paragraph, if we are suffering here from confusion about "the real game being played" ([7], p. 144). I had thought that the 'real game' was to protect patient freedom and autonomy in health care, and to institutionalize the insight that patients are the ones who know best for the simple reason that it is the best *for them* that we are talking about, not for someone else. Is the 'real game' instead one in which we are going to allow the *form* of patient decision-making only as long as the *content* of the decision squares with the opinion of the experts? But this is the old wolf of paternalism and informed consent is just the sheep's clothing — there is no sheep here.

To conclude: my position here is that the new ethos of medicine, seen simply as an extension to health care of what it means to be a member of a free society, cannot countenance the more stringent gate-keeping that some of my colleagues recommend here. One may surely request that health care providers attempt to respond therapeutically to competence-diminishing factors, as well as to assist patients to come to the best understanding of their situation and options they are capable of. Equally, one might argue that such factors be given more weight in judgments of incompetency. Further, there is the whole area of the ongoing physician-patient relationship wherein we respond to the three clinical realities with advice, education, therapy, and counseling within the bounds of a general respect for, and ultimate acquiescence to, the patient's autonomy. I submit, in fact, that our 'clinical realities' supply strong grounds for attending to this 'murky' area in much greater detail in lieu of trying to generate legal maneuvers that jeopardize the freedom of individuals. We may even find that the legal doctrine of informed consent itself stems as well from a 'to the task' competence sort of base, and that we would be better served by seeking to enhance (as well as trust) the dynamics of the relationship between the patient and a conscientious physician ([10], pp. 272—273). But, regarding our concerns here and now, if the patient retains enough capacity to merit inclusion into the class of general competence, then his or her decisions and directives must be heeded, *not* placed under special burdens as if we really do not believe in freedom or the moral restraint it requires.

State University of New York
Buffalo, New York

BIBLIOGRAPHY

1. Beauchamp, T. L.: 1991, 'Competent Judgments of the Competence to Consent', in this volume, pp. 49—77.
2. Knight, J. A.: 1991, 'Judging Competence: When the Psychiatrist Need, and Need Not, Be Involved', in this volume, pp. 3—28.
3. Morreim, E. H.: 1991, 'Different Notions of Autonomy and Competence: Their Implications for Medicine and Public Policy', in this volume, pp. 93—125.
4. Nozick, R.: 1974, *Anarchy, State and Utopia*, Basic Books, New York.
5. Pellegrino, E. D.: 1985, 'Informed Judgments of Incompetence; The Patient, The Family and The Physician', in this volume, pp. 25—45.
6. Pincoffs, E.: 1991, 'Judgments of Incompetence and Their Moral Presuppositions', in this volume, pp. 79—89.
7. Robertson, J. R.: 1991, 'The Geography of Competency', in this volume, pp. 127—148.
8. Robinson, G. and Merav, A.: 1976, 'Informed Consent: Recall by Patients Tested Postoperatively', *The Annals of Thoracic Surgery* **22**, 209—212.
9. Roth, L. *et al.*: 1977, 'Tests of Competency to Consent to Treatment', *The American Journal of Psychiatry* **134**, 279—283.
10. Wear, S.: 1983, 'Patient Autonomy, Paternalism and the Conscientious Physician', *Theoretical Medicine* **4**, 253—274.
11. Weinstein, M. *et al.*: 1980, *Clinical Decision Analysis*, W. B. Saunders, Philadelphia.

BREAKING UP THE SHELL GAME OF CONSEQUENTIALISM: INCOMPETENCE — CONCEPT AND ETHICS

Should a man faced with the dreadful diagnosis of throat cancer be permitted to refuse an operation (a radical neck dissection) thought to be best life-prolonging treatment for his condition [13]? Should an elderly inmate of a psychiatric hospital be permitted to refuse a biopsy and follow-up care of a lump in her breast where cancer is suspected [12]? Should a middle-aged man with severe headache and high fever be permitted to refuse care and die when diagnosed to have bacterial meningitis, which is easily treated by antibiotics [6]? Should a bitter hemophiliac be permitted to refuse the care of a woman surgeon just because he is a misogynist [11]?

To some extent, the answers to these questions depend on whether the patient in question is competent to make the decision in question. This reply, however, is ambiguous. For 'competence' and its opposite, 'incompetence', are sometimes taken to be purely legal terms in that, according to the requirements of due process of law, only a court is authorized to declare someone incompetent and deprive that person, that citizen, of the exercise of his or her rights and choices. Sometimes, though, the terms are taken to be mental terms and seem to refer to the patient's states of mind and mental abilities — whether his beliefs are reasonable, whether his memory is adequate, whether his powers of inference are sound.

When this is the context, the assessment of competence may be easy. It is easy in cases when patients are newborn babies or unconscious accident victims. Such voiceless individuals force the responsibility for decisions about their welfare upon others. But when a patient requests that his doctors follow courses of action different from those thought to be to the patient's welfare, the question of competence can arise in a more difficult way. It is difficult, it seems to me, for the following three reasons. First, there is conflict with culturally powerful experts; second, a great deal can turn on who has his way; and finally, knowing another's mind is in itself sometimes very challenging.

We should make special note of these three ingredients of difficulty.

The experts have been trained to help with matters that lay persons cannot be expected to manage, and the experts have come to expect to have control over the decisions in their area of expertise. They have also assumed a set of values about and duties to life and other areas of patients' bodily welfare. However, society's emerging recognition of patients' rights has created dilemmas for the experts, because patient autonomy means that patients can refuse courses of action preferred by the experts for the patient's benefit. Declaring a patient incompetent is one way to attempt to legitimate overriding the patient's statement of refusal.

But when, why, and how should the experts bother to attempt this disenfranchisement? And more to the point: when should they succeed? Perhaps they should attempt it when patients reject what the doctors want to do *and* complying with that rejection is thought to be very serious. This seems to be when and often why they attempt patient disenfranchisement. So these conditions seem to be necessary contextual conditions or triggers for the attempt. Sometimes, however, physicians do not feel comfortable making the declaration and attempt on their own. They seem to want validation from someone else, typically a psychiatrist. This desire can be taken to stem from the insight that the proper declaration of incompetence will turn on the assessment of the mental abilities of the person in question. This is the third ingredient of the difficulty in assessing competence, and it raises the question: what tests should be used to assess the patient's mental abilities?

We want the criteria to sort those who have defects of reason for the task in question from those who do not. As Pincoffs has argued, if we were designing a society for ourselves, but did not know the exact place each of us would hold in it once it was initiated, it would be a part of the design that we do not automatically turn ourselves over to doctors to do what they think best [16]. The fairness and authenticity of the competency tests are required to protect us from turning that intended aspect of the society's design into a sham. Accordingly, we need criteria of incompetence that can be applied objectively by persons who are not parties to the conflicts patients have with doctors about what is in the patient's best interests. As an ideal, that objective application would be the duty of the psychiatric consultant and, later, of the court. (In practice, as we know, the court is frequently left out; somewhat less frequently the psychiatrist is, too.)

Regarding these criteria, something of a conceptual dilemma has emerged. There are good reasons to hold that incompetence should not be an all or nothing status [13]. Someone might be incompetent to marry, but be competent to make a medical decision. Thus, it would seem improper to make global assessments of patients who have some mental capacity. If we are limited in the scope of abilities we are entitled to assess, how can we make an assessment other than by judging the consequences of what the patient will undergo? If that is to be the focus, we are right back with deferring to the experts' opinion about what is best for the patient, and the test will be a sham. That is, either we must import standards to hold the patient to from contexts and problems other than those he or she currently faces, or we must use outcome assessment in deciding about competence and incompetence. Each choice has gains and losses. To use imported standards is to run the risk of using the wrong ones and letting someone decide who lacks the ability to do it well. But using these standards has the virtue of respecting the patient as a chooser, while allowing some unhappy outcomes. To use outcome assessments as the standard of competence, however, is to return to the experts the power that our ideal social design was supposed to preserve for us as free and equally enfranchised agents.

With this dilemma in mind, we can assess the seven tests of incompetence that Beauchamp listed and considered — surprisingly making it appear arbitrary which test should be employed. The seven tests involve (from easiest to most difficult to pass) the ability (1) to indicate a preference, (2) to understand one's situation, (3) to understand provided information, (4) to give *any* reason for the choice, (5) to give a *rational* reason for the choice, (6) to give risk/benefit reasons for the choice, and (7) to reach a reasonable decision in terms of the expectable outcome of the choice.

Beauchamp tells us that selecting which to employ in a given case is complex and depends upon personal values of the tester. If the tester wants most to protect autonomy, he or she will use one of the first three. To protect welfare, one would employ test (6) or (7).

Beauchamp characterizes the concepts *autonomy, beneficience*, and (by implication) *competence* as essentially contested concepts. A concept is essentially contested when no criteria for its application are considered to be necessary conditions of its applicability and no clear threshold exists for when groups of the criteria make it definitively applicable. Regarding competence, this means that the physician must weigh respect for autonomy against concern for the patient's welfare and then decide competency based on that. Beauchamp seems to be saying *first* that the *choice* of tests is a case-by-case decision made from an evaluative point of view. The evaluation involves the issue about whether to emphasize esteem for autonomy and liberty or medical outcome. In short, the choice of tests is not empirical. But *second*, once a test has been chosen, he believes that its *application* should be done from an objective, empirical viewpoint.

I agree that the choice of tests is no empirical matter. However, I do not think that the choice should be made on a case-by-case basis or that the evaluative perspective should vary according to whom is choosing and how much she esteems liberty versus how much she esteems welfare. Beauchamp's position on defining competence and establishing criteria and tests for it is haunted by the absurdities of the problem of other minds [14]. It commits a logical absurdity, *and* it lends itself to moral abuse. Beauchamp and others ([1], [2]) accept the thesis that competence will vary by which test one adopts, given the values one wants to preserve (autonomy or welfare). This makes the mental realities unreal and is too disrespectful to personhood. People who argue that way are guilty of a sort of behavioristic obverse of the fallacies in the doctrine of the unknowability of other minds. In the original statement of the problem, the truth is declared unknowable because it is mental and therefore intrinsically inaccessible to observation. In the sort of value-decision position held by Beauchamp, there is no truth about the patient's ability and the tests chosen are not arbitrary unless they are about outcome. Ignoring or denying the patient's mental life allows such thinkers to hold that outcomes should be determinative of whether the patient be considered competent. The Beauchamp-like position may well describe what is done — the issue is: may we logically and morally do what sociologically is found?

An essay by Benjamin Freedman [10] convinces me that we should not conduct ourselves according to the amalgam that includes Beauchamp's position. Following Freedman's logic (he addressed somewhat different tests), we may say that of the tests that Beauchamp identifies — (1) ability to express a preference, and (7) not opting for a better outcome — are unacceptable extremes. Each destroys competence as a notion by denying all interest in any knowledge of the particular patient's ability to think.[1] If passing the first is taken as sufficient, we are accepting any behavior reminiscent of drawing a conclusion, with no regard for reason. Using the seventh is accepting any course considered by professionals to be the least morbid — again, regardless of the patient's reasons. Neither of these tests requires more knowledge about the patient, and neither is designed to protect the freedom of those who are capable of being respected as free persons even in the narrow context of making a particular decision.

Tests (2) and (3) capture an aspect or necessary condition of competence [as does (1)] but neither should be considered sufficient. For, competence requires having reasons. The question is what kind of reasons?

In the context of our interest in competence, the understanding of a situation [test (2)] and the understanding of information [test (3)] cannot logically be separated from the ability to use that information. Thus, we must choose from among tests (4) any reasons, (5) rational reasons, and (6) risk/benefit reasons. But surely counting the ability to give any reason for a choice (test 4) as a sufficient condition of competence is as muddled as accepting any gesture behaviorally reminiscent of a preference [test (1)]. Accepting, for purposes of determining competence, *anything* offered as a reason destroys the meaning of the term 'reason'. Indeed, we are interested in the competence of a patient because we wish to know whether she reasons well about her situation and choices.

We are left, then, with deciding whether the reasons have to be compelling in a cost/benefit or risk/benefit mode (test 6) or whether recognizable or rational reasons (test 5) should be the standard. As a paradigm, risk/benefit decision-making requires probability statements about outcomes multiplied by quantifications of morbidities and the like. In the context we are focused upon, it would seem impossible to have such statements about each individual patient. Indeed, I believe that almost all risk or cost/benefit talk, *even on the part of the professional staff*, is metaphoric. Which morbid outcome should be weighted at .7 and which at .6 is subjective and arbitrary. Casting qualitative expectations about a person's welfare as definite, specific numbers make such risk/benefit calculation appear more precise and definitive than it can be. It also ignores the different weights that different persons would give different suffering as well as ignores many of their individual interests.

Thus, what Freedman calls "recognizable reasons" ([10], p. 64) should reign. These are reasons and arguments that we may not find decisive but must agree are relevant. If called "rational reasons" [Beauchamp's test (5)], the name of such a test may seem self-serving, and surely it is subject to some of the rough spots identified by Beauchamp in his fifth footnote. But it is not without important applicability, and I believe its richness is yet to be appreciated (even by Freedman, as my differences with him on *In re Yetter* will show).

Freedman contends that for an argument to be a recognizable reason it should have accurate premises and a conclusion related to those premises in a strong, but not necessarily valid way. He probably overstated the case. Requiring accurate premises is too stringent. There are types of premises including general medical, particular medical, and non-medical ones. Thus, someone may believe that all medications are dangerous or that the specific ones being prescribed are or that if he takes them the Martians will invade. If the general empirical contextual premises are wildly inaccurate, we may not find the argument reasonable. "That there are Martians and that they are ready to invade" could not form part of the grounds of allowing a patient to refuse or accept treatment.

Because this may be the knottiest problem in the set of issues we are considering, perhaps the clearest we can be about premises is this. We should try to inform patients of our disagreement with their intelligible and plausibly grounded premises, and we should tolerate their use of such premises about what will happen to them [9]. This, together with our willingness to accept a "strong, but not valid" relationship between premises and conclusion allows us to respect those with whom we may disagree. For, after we have tried to engage them about their premises and the invalidity of their reasoning, we must accept them for the persons they are (a point to which I return). Reasoning that is "strong, but not valid" probably is no looser, I would add, than most of what passes as risk/benefit reasoning. And though this phrasing acknowledges a high sense of uncertainty in assessing competence, it lets us express respect for personal integrity, in contrast to fixation upon consequences. We must now refute the appeal to consequences.

Quite near the conclusion of his essay Beauchamp declares, "Just as requirements for obtaining informed consent should be increased as the risks and related possible consequences of the procedure increase . . . , so the standards of competence do and should increase with the consequences of the affected party" ([3], p. 71). Even closer to the end he writes, "there is no such thing as competence apart from our *moral* judgments about where appropriate thresholds *should* be set" (his italics) ([3], p. 71). This allows too much of a semantic shell game.[2] And it runs grave risks of harming or wronging persons and indulging the professionals' teste for control.

We should reject the position of authors who define meaning on a case basis determined consequentially. According to Freedman, such authors seem

> to be saying that rationality multiplied by harm equals competence, while we would have thought rationality itself equals competence. For what difference does it make that a choice is serious, provided that it can be comprehended by the individual? ([10], p. 67).

Two further points of Freedman's are important here. First, we should not confuse concept and policy. *Rationality* and *competence* are definable independent of consequences. Regarding policy, the best approach may require more vigilance in determining rationality when the consequences of making a choice are very serious regarding the welfare of patients. Regarding concepts, let us keep clear that we mean the mental abilities of patients.

Second, a 'play it safe' attitude or practice is morally dangerous and conceptually confused. I put it rhetorically thus:

> If a patient refuses the option which his physician considers best, and there are *any* (other?) grounds for doubting his competence, then isn't it safest to deny this competence and perform the procedure? If one choice doesn't harm him or if it even benefits him but he has chosen a harmful option, forcing the harmless/beneficial one is the morally preferable course.

Such thinking "may really be neglecting a less obvious, but for all that no less serious, harm" ([10], p. 68). Freedman rightly argues that it is a serious harm and an injurious insult to personality wrongly to judge someone incompetent and treat him in that way. Freedman declares:

> It is easy to ignore this relatively inchoate harm, or to see it overshadowed in the light of the measurable physical harm that can result [for example] from a failure to amputate. Yet for this very reason it is important to insist upon claims of the psyche. There is rarely any way of playing it safe in this world. The best we can hope for is clear analysis and good will ([10], p. 69).

The point, then, is to avoid developing or employing a policy designed to protect patients rather than to reveal their rationality. *Allow protection and safety their place in how carefully competence is tested — test most vigorously when the choice can have serious consequences; but do this as a matter of policy, not definition.*[3]

A corellary to the notion of respecting the psyche is the relationship between autonomy and competence. At the end of Part II of his essay, Beauchamp declares that autonomy is a criterion of competence and then defines autonomy as freedom together with the ability to comprehend information and to act intentionally. Mention of the ability to comprehend makes his analysis seem viciously circular because the ability to understand is part of what is at issue in deciding competence.

Following Bruce Miller in large degree, it seems to me more fruitful to worry about whether a patient's consent or refusal is adequate to her status as a self [15]. Accordingly, we should be concerned about whether what she expresses as a choice is (1) *voluntary* — meaning we can rule out coercion. We should also be concerned about whether her choice is (2) *authentic* — including not voiced out of having been deceived or out of delirium or perverse desperation due, e.g., to an angry reaction to poor communication from the physicians [2]. We hope to be sure from this that an avoidable set of causes is not diverting her from choosing just what she wants from the range of available and fair options. In short, we are hoping to foster the patient's choosing on the basis of values she uses to define herself.

We could, further, be concerned about (3) whether the patient's choice is *reflective*; that is, whether it involves values that have been deliberately adopted by the patient. Does the patient accept as her own the values from which the choice is derived? Has she gone through a Socratic rebirth and given authority to her values? This, of course, is a very demanding criterion. It is only brushed in cases where consent is refused (for example, when a doctor asks: "Are you sure that is what you want to do?" or a nurse says "Think of how your family will suffer without you"). This aspect of autonomy may be too difficult to assess or assessing it may be too invasive of the self for it to have much to do with acceptable practices of assessing competence.

The remaining sense of 'autonomy' on Miller's list is (4) *effective deliberation*. This is what we most typically mean by 'competence' but should be speaking of in terms of 'rationality'. It includes that the patient recognizes that she is in a situation calling for a decision, knows the alternatives open to her, understands the consequences (harms, risks and benefits) of each. Notions that contrast with this item include impulsiveness and irrationality.

There are circumstances where persons have been willing to override on moral grounds an unimpaired person's fully autonomous choice. Economic arrangements may incline some to withhold costly treatments from autonomous competent people (*cf.* England's restrictions tied to age). Regard for symbolic or spiritual dimensions between people may (as in Ms. Bouvia's attempts to force a hospital to assist her suicide by starvation) incline a judge to override a choice, for the sake of the integrity of the institution of medicine.

Contrariwise, there are cases where someone might well be judged incompetent and yet the toll on her of overriding her refusal of treatment might be too great for us to accept morally. *In re Yetter* [12] offers an example of a case like this. As a committed mental patient in her sixties diagnosed as schizophrenic many years before, Ms. Yetter refused surgery on her breast to test for or to remove cancer. First, she refused it out of fear of death from the operation. She was non-delusional, and understood the risks of foregoing treatment. Months later, her reasons included claims that the operation would compromise her genital system, diminish her ability to have babies, and destroy her chances of a movie career.

The courts were approached. Ms. Yetter's composure during her hearing was exemplary. Judge Williams decided in favor of her expressed desire even while acknowledging the decision as irrational! He declared that in general an "unwise, foolish or ridiculous" but competently made decision should be respected and that in Ms. Yetter's case that description fit her *original* decision. Her elaborations, though making the decision seem irrational, were ruled either acceptable or irrelevant.

Now I believe that the judge ruled well[4], save for the affirmation that she was competent to make this decision. If Beauchamp's position were to hold on this point, we would have to choose whether to set the test of competence very low and side with freedom or to set it high and side with safety. But, to go beyond Beauchamp, I believe we should find her irrational and incompetent on this issue but find ourselves obliged to go along with her, for reasons suggested by Miller's analysis. To have her undergo the operation, persons would have to force her. She would experience being a victim of violence. She would likely hate her body, her physicians and staff, her future, and lose whatever trust she might have. In short, her interests seem to have consolidated enough into a personality that even though her premises are not sound, the consequences of not respecting her psyche are unconscionable (a point also made by Robertson ([17]).

All these examples are meant to show that, as a matter of social practice, we use three sets of moral ideas to make moral assessments: (1) *welfare* — as defined by scientific criteria, (2) *interests* — as defined by the person's sense of self, and (3) *social ties* — which include the patient's legal rights and the virtue of living in a society governed by the rule of law [8]. We can use these to justify (with varying levels of convincingness) both overriding autonomous, competent decisions that request treatment and overriding autonomous, competent decisions that refuse treatment. We may also morally respect incompetently made decisions. If not, we have to accept the difficult idea that, given the sense of self that someone like Ms. Yetter achieves, she and others like her are 'task competent' to decide what fits her interests and sense of self. (Again, see Robertson [17].)

This approach cannot *settle* all general theoretical problems about how the incompetent or marginally competent patient should be managed. Also, we must remember that competence is neither morally necessary nor morally sufficient for conforming with a patient's choice. It is a *prima facie* criterion, one that is very important and should, as a matter of policy, only be overruled when the values at stake have been carefully identified, fairly 'weighed' and fully 'aired', and when the mental skills and sense of self of the patient have been well explored. Beyond the typical but vast difficulties of doing this ([5] pp. 381–383), there are uncommunicative, defiant patients, who will not cooperate, and there are highly ambivalent, dependent persons who cannot bring themselves to make a decision. Some delay should be tolerated by hospitals and staff rather than giving in to the temptation to use some sliding scale of competence to effectuate a decision and appear to be innocent of unwarranted paternalism. Intellectual honesty requires that we not use the sliding scale as part of our *concept*, though we may well use a scale to probe into a patient's performance.

It remains to discuss those cases mentioned in the first paragraph that have not yet been discussed. The first case listed is taken from Knight's discussion [13]. His account confirms much of what I claimed in the earliest part of my essay. It was conflict with authority, rejection of the course of action recommended by experts that gave rise to the surgeons questioning the patient's competence in the first place. On the basis of his rejecting what they considered the course of action least likely to have worst outcome, they believed that he must be crazy and sought psychiatric confirmation. One could ask why they sought the consultation at all. With an unconscious patient, they would not have had a conflict, and possibly would have had their way. Although they expected psychiatric confirmation, they sensed that they needed it, perhaps for legal reasons. But the patient was able to communicate enough and in sufficient style that they also seem vaguely to have sensed that not merely conflict with them and a grave outcome would be enough to gain a declaration of incompetence. Their surprise teaches us about their presumptions and presuppositions, and it shows that declaring someone mentally incompetent need not be automatic and need not be a sham.

Consider now the case of the man with meningitis from the opening paragraph of this essay. Eric Cassell, in discussing this case, hopes that a clinic doctor would call for a psychiatric consultation to declare the patient incompetent. This would enable the doctor to be safe from legal liability and treat the patient [6].

A rights-oriented reader might be offended by Cassell's argument. But I believe the offense would be due more to its structure than its content. For, at first, Cassell tries to justify the intervention by claiming that doctors get nearly blanket permission to save lives the moment that persons come to them as patients! He believes that this patient is attempting to force the doctor to assist in the patient's suicide. Cassell writes, "A patient who refused further artificial kidney dialysis could be declared incompetent on the basis of the fact that his refusal constituted suicide" ([6], pp. 16–17), and argues that the refusal would be honored because the costs of incarcerating the patient and coercing him to put up with treatment three times a week would not seem reasonable. In this phase of his essay, Cassell seems wrong, even histrionic, if not hysterical. But in this phase he has only been disclosing his motives.

Cassell's discussion of reasons is much better. For, clearly, being sick can compromise judgment, and it is very likely that this patient's judgment is impaired. These are good reasons to doubt the authenticity of the patient's request to be allowed to die. This is just what Pincoffs' analysis of the role of mental terms should lead us to suspect [16]. The pain and fever are good grounds to exclude the idea that the person is acting autonomously. The signs and symptoms constitute grounds to declare the patient incompetent because they are likely to be affecting his ability to think. We do not have to concern ourselves with consequences at *this* juncture. For here we are analyzing the concept. When we turn to policy, however, the concern for the seriousness of the consequences should make us probe his reasons to see whether they are recognizable and sincere. Along these lines, we would be using the right test in the right way. If we do not have time to probe the patient's reasons or if he will not disclose or discuss them with us, maybe then concern for consequence to him could lead us to risk damaging his psyche and to endorse treating him. We could do this paternalistically. Or, because the disease is highly contagious and dangerous, we could do this on ground of public health alone; and, in this case, leave the concept and policies about incompetence unabused.

Some of the implications of these ideas about intellectual honesty are obvious and sad. We may, *pace* Cassell, feel obliged in some cases to let people die or become sicker or suffer deformity when these consequences could possibly be avoided. We may, also, have to accept that in some cases we have curtailed a competent person's liberty for the sake of some other value (Ms. Bouvia and the integrity of medicine). If we do such things, we should not make ourselves feel better by conceptual slight of mind or semantic trickery, but identify what we are doing as honestly as possible, justify it as fully as possible and assess the justifications as best as we can. In this way, we can avoid the harm done to a psyche if we assume that a particular choice against a medical intervention is irrational or incompetent. But other potential abuses of psyche by way of uses of the policies surrounding incompetence still can be found.

Consider the last case I will discuss in this essay. It comes from the *Massachusetts General Hospital Handbook of General Psychiatry*. The case description identifies Dr. B as a brilliant and dedicated surgeon who was faced by a Mr. J who refused to allow her to examine him because he dislikes women. As a hemophiliac, he had a grudge against all women because they carry the disease but cannot get it. Mr. J had been hospitalized for a hernia with possible incarcerated bowel. We are told that Dr. B "was strongly reluctant to place him in the care of a male physician on another team, but was concerned about what might occur if his (Mr. J's) condition worsened" ([11], p. 555). We are not told *what* Dr. B is most concerned with — whether patient welfare *or* her own legal liability arising from her wanting to treat him without his consent, an act that would render her liable to a battery suit.[5] But it does seem that ego and haughty professionalism more than optimal clinical care are on Dr. B's mind. At any rate, the consultant helped the patient by teaching him to perform the test which the patient most feared. The consultant helped the doctor by informing the patient that although he was sane and competent ("not insane and not incompetent" is the grudging phrase used on page 556) "the hospital was not obliged to provide a male physician in such a nonemergency situation. He told the patient, in short, that he might have to choose between the female surgeon and no physician at all ... He ... told ... [the doctor] that she was not obligated to transfer the patient nor liable if she did not" ([11], p. 556).

I cannot assess the psychiatrist's claim about liability. But it seems clear that best clinical care is not top on the list of concerns. Control, power, etc., seem higher. Further, no one is treating this patient as the person he is — a point to which I shall return. If the hospital has no (legal) obligation to supply a male physician to a patient in a nonemergency situation, surely its duties along these lines in a true emergency are less since often consent can be presumed, and refusal of lifesaving treatment on grounds of a doctor's sex probably would not register as a recognizable reason. In this case, however, one wonders what would have happened if the psychiatrist also had been a woman and a feminist! What 'findings' would be more likely? And in this circumstance would the psychiatrist have pushed the patient to leave against medical advice under the formula 'Dr. B or no one'?

This case discussion occurs in a section of the *Handbook* that focuses on the ('rare') conflicts and contradictions between law and medicine. The section is entitled, 'Theory versus Practice: Law and Medicine'. One should wonder *which* conflicts and contradictions the authors have in mind. What does the law say or what do they think it says? And does 'medicine' in this subtitle mean "the enterprise of rendering scientifically based assistance" or is it rather "the institutional social values of the people who give the assistance"? What personal and professional gut values and notions of practicality are at work here? How do they override the (personal and) professional ideals involving the importance of attitude and the quality of emotions in the doctor-patient relationship?

The discussion of legal 'theory' (impracticality?) versus medical 'practicality' (realism?) is in the section of the *Handbook* immediately following the discussion of competence. Its very title, place and style seem intended to ridicule the law or to teach how to circumvent it (and thereby how to circumvent patient choice and freedom). The section seems to be about competence and incompetence as tools for manipulating patients for the doctor's non-medical desires. The position it states is more alarming than that taken by the doctors who wanted to override a patient's refusal of an operation to remove throat cancer. In the case from MGH, it seems that the doctors wanted to overcome the personal values of the patient for a doctor's sake — with no regard for patient welfare or interests or rights. This is a disheartening and surprising implication that leads, I believe, to yet a further disheartening surprise.

The surprise I have in mind is this. As humans emerge into individual persons with self-concepts and world views, some (perhaps those considered borderline personalities) become petty, angry, begrudging, inconstant individuals.[6] If we are to accept the implications of treating them as persons, we should, I propose, accept them as they are (or as we think we have found them). We should indulge their prejudices within the scope of law and public policy. And, in dealing with them in situations (not the case of Dr. B and Mr. J) where one might be inclined to say of someone that he needs to be protected from himself, we are in a bind. We must remain aware that this kind of paternalism does not respect psyche, personality, or privacy. It is more an invasion of self than ordinary paternalism, for it does not merely disregard autonomy, it disregards the whole person's avowed inclinations.

In such circumstances, the destructive skepticism about other minds that produces problems can remind us of an insight. It is sometimes very difficult to know another self! How presumptuous it is, then, to protect a self from itself! How tempting it is to treat someone we dislike as incompetent! At least the psychiatrist's negotiating between Mr. J and Dr. B avoided those pitfalls. But it might have been a close pitfall, if we can, as Beauchamp has urged, use a sliding scale of values and criteria to determine competence. The very presence of a psychiatrist assessing a person for determinations of competence can be an invasion of freedom and self.

We can surely judge competence by the person's performance on closely associated tasks and by the meaning his experiences will have for him (even if some of his important premises are false), rather than by putting welfare-oriented consequences, by slight of hand, into the meaning of a mental term. We resist obscuring the notion of competence by not allowing the test for it to be a function of whether the person who chooses the test fancies autonomy more or welfare more.

As an ideal, the choice of a course of action for a medical problem is a choice that is voluntary, authentic, reflective and the product of effective deliberation based on adequate information. Unfortunately, many choices fall short of the ideal. Even forcing a person to 'share' his/her value orientation and thinking with members of the health care team could be a moral violation of personhood and privacy. In the medical context, action has a very high value, as we know. Decisiveness does, also. So, too, does patient compliance.

As the DRGs and related systems evolve, they will encourage decision-making and earlier discharge from hospitals. The quality of the processes of disclosing information and getting authentic, voluntary, reflective decisions will likely suffer. Perhaps refusal will be taken too glibly ([2], [17]). Perhaps incompetence will be declared too easily. It is vital to resist those aspects of change. For freedom has many aspects about it that make it a process rather than a state [10]. And making serious life's choices on the basis of the time-table of others is very compromising.

Rationality is a *necessary* condition of liberty and having rights. Competence is neither a necessary nor a sufficient condition, for an adequate rights claim, because other social ties may block a person's free choice, and we might want to comply with the demands of a self that is not normally rational but is somewhat integrated. We may also find self-destructive action to be grounds to justify blocking a liberty. But it would never do to declare the self-destructive person (patient?) incompetent on that basis alone.[7]

University of Medicine and Dentistry of New Jersey
School of Osteopathic Medicine
Stratford, New Jersey

NOTES

[1] Being unable to communicate may be sufficient to declare someone incompetent but surely is no proof of mental inability. This arrays well an important difference between a social (legal) status and a mental one.

[2] Even requiring that a reason be given may commit moral wrongs, since many people might be competent and unwilling to give any or honest reasons.

[3] James F. Drane [7] conflates concept and policy, but offers a fairly good policy. For example, in serious acute illnesses which are easily treated, a consent can be taken to be competently given and the patient need not be looked at closely to determine competence. But this doesn't change the meaning or criteria of competence.

[4] In arguing this, I am in stark conflict with Freedman's discussion of this case ([10], p.

252   EDMUND L. ERDE

62), but I think this is due to our *use* of distinctions, not a disagreement about the distinctions themselves.

⁵ The legal concern is more plausible since the entire section in which the case is discussed is entitled, 'Theory Versus Practice: Law and Medicine'. It is about apparent *contradictions and conflicts* "between optimal clinical care of a patient and the law" ([11], p. 555).

⁶ For an interesting discussion of this type of person, see Frithjof Bergmann's discussion of Dostoevsky's *Underground Man* [4].

⁷ I wish to thank Benjamin Freedman and James B. Speer for comments on an early draft of this essay. They helped improve it a great deal. Flaws, of course, are my responsibility.

BIBLIOGRAPHY

1. Appelbaum, P. S. and Roth L. H.: 1982, 'Competency to Consent to Research', *Archives of General Psychiatry* **39**, 951—958.
2. Appelbaum, P. S. and Roth L. H.: 1983, 'Patients Who Refuse Treatment in Medical Hospitals', *Journal of the American Medical Association* **250**, 1296—1301.
3. Beauchamp, T.: 1991, 'Competent Judgments of the Competence to Consent', in this volume, pp. 49—77.
4. Bergmann, F.: 1977, *On Being Free*, University of Notre Dame Press, South Bend.
5. Buchanan, A.: 1978, 'Medical Paternalism', *Philosophy and Public Affairs* **4**, 370—390.
6. Cassell, E.: 1977, 'The Function of Medicine', *Hastings Center Report* **7**, 16—19.
7. Drane, J. F.: 1984, 'Competency to Give an Informed Consent', *Journal of the American Medical Association* **252**, 925—927.
8. Erde, E. L.: 1985, 'The Virtue(s) of Medicine', in E. Shelp (ed.), *Virtue and Medicine*, D. Reidel Publishing Company, Dordrecht, Holland, pp. 201—222.
9. Faden, R. and Faden, A.: 1977, 'False Belief and the Refusal of Medical Treatment', *Journal of Medical Ethics* **3**, 133—136.
10. Freedman, B,: 1981, 'Competence, Marginal and Otherwise', *International Journal of Law and Psychiatry* **4**, 53—72.
11. Hackett, T. P. and N. H. Cassem: 1978, *Massachusetts General Handbook of General Psychiatry*, The C. V. Mosby Co., St. Louis.
12. In re Yetter, 62 Dec. 2d (C. P. Northampton City, P.A. 1973).
13. Knight, J. A.: 1991, 'Judging Competence: When the Psychiatrist Need, or Need Not, Be Involved', in this volume, pp. 3—28.
14. Malcolm, N.: 'Wittgenstein's *Philosophical Investigations*', in N. Malcolm (ed.), *Knowledge and Certainty*, Prentice-Hall, Engelwood Cliffs, New Jersey.
15. Miller, B.: 1981, 'Autonomy and the Refusal of Lifesaving Treatment', *Hastings Center Report* **11**, 22—28.
16. Pincoffs, E.: 1991, 'Judgments of Incompetence and Their Moral Presuppositions', in this volume, pp. 79—89.
17. Robertson, J.: 1991, 'The Geography of Competency', in this volume, pp. 127—148.

COMPETENCY: A TRIAXIAL CONCEPT

I. INTRODUCTION

The notion of 'competency' (in the sense of 'mental competency' or 'decision-making capacity') lies at the intersection of a number of important issues in contemporary medical ethics.[1] 'Competency' in its general sense is usually understood to be a precondition to the exercise of autonomous choice. An individual who is determined to be 'incompetent' may not be permitted to act in a way he perceives to be in his best interests, including the accepting and refusing of medical diagnostic and therapeutic procedures. Thus, a person who is in fact competent but found to be 'incompetent' may lose his right to self-determination. On the other hand, an individual who *is in fact incompetent*, may so act as to bring harm to himself; such a person is not functioning 'autonomously' and protecting such an individual can be justified on the basis of a principle of beneficence.[2] Though a determination of 'incompetency' in its strict sense is a *legal* determination and is made by a judge, the determination typically rests on information and an interpretation of that information provided by medical personnel and in particular by psychiatrists. More significantly, many determinations of competency/incompetency that play important parts in patient management are made by attending physicians who are not psychiatrists.[3] Thus not only ethical, legal, psychiatric, and social issues are involved in competency determinations, but also *medical* management decisions arise in these situations.

Ethical and legal issues involved in determinations of competency are likely to continue to be of significant interest in the area of medical ethics. The determination often arises in connection with decisions to treat or not to treat a patient, especially when the patient, the patient's family, and the attending physician's decisions are in opposition.[4] Such decisions are being made certainly more visibly and probably more frequently ([26], pp. iii—iv). Competency determinations are likely to become of increasing significance as the prevalence of Alzheimer's disease continues to increase, and will almost certainly be involved in

253

the increasing debate about 'rational suicide' for terminally ill patients.[5] Though, as Robertson astutely points out [41], a competency determination does not produce *closure* on patient management decisions, I believe that a detailed analysis of the concept of competency and its test should assist us in decision-making in these situations.

In the past decade, interest in maximizing patient self-determination and a focus on both the ethical and legal aspects of the notion of 'informed consent' began to provoke concern with the concept of 'competency'. But as early as 1941, Professor Milton D. Green analyzed various legal tests found in a variety of judicial decisions and wrote:

> ... if one reads the cases critically it will be found that no verbal formulation of a test can be made which will fit the standards laid down by the courts. So diverse is the phraseology of the test by courts in different jurisdictions, and even by various opinions in the same jurisdiction, that no single statement of a rule can be constructed which, if it has meaning, will not exclude a majority of the cases ([21], p. 147).

More recent commentators have noted a similar range of tests and meanings of the notion of competency. In the following section I will review some of these recent accounts in preparation for an analysis of 'competency' which I will argue captures somewhat more of the core features of this extraordinarily important but still poorly understood notion.

II. RECENT APPROACHES TO THE ANALYSIS OF COMPETENCY

In this section, I will discuss three recent approaches to the concept of mental competency. I begin with the work of Roth, Meisel, and Lidz, later joined by Appelbaum. I then discuss the approach of Gert and Culver, and also consider the recommendations of the President's Commission for the Study of Ethical Problems in Medicine and Biomedical and Behavioral Research (President's Commission). Finally, I comment on several other authors whose work, though not in my view representing fully developed accounts of competency, nevertheless makes important points which I think any reasonable analysis of competency should consider.

Roth, Meisel, Lidz, and Appelbaum

In 1977, Roth, Meisel, and Lidz, motivated in part by their then emerging research project on informed consent, wrote that:

... competency is theoretically one of the independent variables that is determinative in part of the legal validity of a patient's consent to or refusal of treatment. There is therefore a need to specify how competency can be determined ([44], p. 280).

In their 1977 article, these authors outlined five tests in an order of increasing strictness:
(1) evidencing a choice
(2) 'reasonable' outcome of choice
(3) choice based on rational reasons
(4) the ability to understand, and
(5) actual understanding.

In subsequent articles, Roth writing with Appelbaum reclassified and elaborated these five tests into essentially four aspects of patient decision-making:
(1) evidencing a choice
(2) evidencing an understanding of relevant issues
(3) rationally manipulating the relevant information, and
(4) an appreciation of the nature of the situation ([3]).

('Appreciation' is defined to encompass not only factual understanding and rational manipulation of the information, but an additional facility for applying general information to particular circumstances and for comprehending the crucial nature of certain data. "Appreciation", Appelbaum and Roth write, "is taken to be an affective, as well as a cognitive, recognition of the nature of the situation" ([3], p. 955).) Again, this numerical sequence is in an order of increasing strictness of test. (In still other work, Meisel and Roth distinguished seven tests of competency which Beauchamp and McCullough believe can be classified under four categories.[6])

Culver and Gert

A somewhat different and influential analysis of competency also appeared during this period based on the joint work of Culver and Gert. These authors proposed a more limited and particularized notion of competency, but one that could then be extended in the proper contexts to cover more traditional problems of understanding the nature and implications of acceptance/refusal of medical therapy. In Culver and Gert's view, competency is *like* the ability to do some *particular* task, but has the additional feature that competence presumes an *expected* ability in that situation. Thus their analysis of competence includes the specification of a role(s), a context(s), and the identification of a particular task (or set of tasks). They write:

> We tentatively define incompetence in the following way: a person is incompetent to do x if it is reasonably expected that any person in his position, or any normal adult human being, can do x, and this person cannot (and his inability to do x is not due to a physical disability) ([10], p. 54).

The notion of 'legal incompetence' requires still another feature, namely, a decision that a different person (a guardian) is justified in acting on the person's behalf to do x.

Culver and Gert's proposal has affinities with the evolving legal doctrine of *limited* guardianship in the sense that a person competent in one area, say treatment decisions about himself or herself, may not be competent in another area, say in balancing a checkbook.[7] Their view has also been largely accepted and elaborated by Beauchamp and McCullough in their synthesis view of competency ([7], pp. 117–124), who also incorporate some of Roth *et al.*'s views in their general account of competency. I shall return to Beauchamp and McCullough's synthesis toward the end of this paper.

President's Commission

At about the same time that the Roth *et al.* analysis was maturing and Culver and Gert were advancing their views, the President's Commission also proposed an analysis of competency. This account suggested that the notion could best be understood under the general heading of "decision-making capacity", which required "to a greater or lesser degree":
(1) possession of a set of values and goals,
(2) the ability to communicate and to understand information, and
(3) the ability to reason and to deliberate about one's choices ([39], p. 57).

For the President's Commission, decision-making capacity is itself but one of three important elements underlying *effective patient participation in health care decision-making*, the other two being *voluntariness* and *information*. The issue of how and in what context to best analyze *voluntariness* is an important one in this paper, and I shall return to it again below. Suffice it for the present to note that in the

President's Commission analysis, voluntariness in *not* viewed as a *component of capacity* (or competency), but is a *closely related notion*.

The President's Commission did not articulate a particular standard for decision-making capacity or competency, though they were critical of any standard based either on a simple "evidencing a choice" or on an "outcome" result, stressing that in their view a *functional* approach involving a patient's *reasoning process* was demanded ([39], pp. 169–175).

The Proposals of Cassell, Siegler, Sherlock, and Drane

There are four other authors who have advanced important additional points related to an analysis of competency. Both Cassell and Siegler have explored what factors should be considered by a physician or other responsible individual in determining whether to override the decision of a patient. (Though a physician would not have the legal authority to so override a patient's decision, she might want to consider the factors to be addressed prior to seeking a court's intervention.)

Cassell, writing on the function of medicine, maintained that most acutely ill patients were devoid of "autonomy". Utilizing Gerald Dworkin's [17] notion of autonomy as "authenticity plus independence," Cassell argued that the function of medicine was to restore a patient's autonomy by treating the patient even against the patient's objections. Authenticity, "the true selfness of a person, the degree to which a person's beliefs, ideas, or actions are truly unique despite their source", is impaired by illness. Independence is also compromised by illness by restricting freedom of choice and highlighting uncertainty.

Siegler, writing with Goldblatt in their essay [50] on the same issue, and following up on his 1977 article [49], proposed five "*prima facie* considerations" which should be used by a physician "when faced with an acutely, critically ill patient who declines life-saving treatment".[8] Siegler and Goldbatt proposed that the physician must ascertain whether the patient "retains sufficient intellect and rationality to make this irreversible choice, and second, whether *his choice reflects a 'true will'* or is merely a reaction to the pain, fear, and uncertainty of his critical condition (my emphasis)." Further, (2) the authenticity (over time) of the patient's choice should be considered, as should (3) the nature of the disease in terms of its certainty, treatability, and prognosis. Finally, (4) coexisting diseases and (5) the patient's age should be factors reviewed in such an evaluation.

Another more recent useful analysis of an important feature of competency appears in an account by Sherlock [46] that begins from the concept of informed consent elaborated by Meisel and his co-workers. Sherlock notes that according to Meisel *et al.*, there are *two* central aspects of a valid informed consent: the *informational* component and the *volitional* component. Sherlock suggests that "the problems of competency are focused around these two core components of informed consent", and maintains that "broad category of *volitional incapacities* is one that has received very limited attention, especially as such incapacities relate to the competency of the patient to give informed consent" ([46], pp. 72, 73).[9] I view Sherlock's point as a valid one and will comment on it further in section 4 below.

The last contribution I want to consider is a suggestion developed by Drane in two articles ([12], [13]). Drane proposes that what he terms a "sliding scale model" will be most useful in competency determinations. On this view, a *minimal* competency standard involving simple awareness and assent is required for high benefit/low risk interventions, whereas a *maximal* standard involving appreciation and rational decision is appropriate for refusal of the same type of interventions or for consent to ineffective treatment. Such a *sliding scale*, Drane argues, rescues competency from being too abstract and perhaps even a meaningless concept.

Drane's approach is not entirely novel, and a number of authors have (perhaps more implicitly) articulated a sliding scale type of approach.[10] The explicit character of Drane's proposal is, however, useful for the account to be given below.

III. EXAMPLES OF COMPETENCY DECISIONS

In this section, I want to outline four cases from the literature on competency which I believe raise salient issues for any general account of competency. The first case I consider is the well-known *Kaimowitz* case [29].

Case I: Kaimowitz

In 1972, a John Doe [later identified as Louis Smith], age 36 and a mental patient at Michigan's Ionia State Hospital who had been civilly committed as a sexual psychopath at age 18, was invited to participate in a state-funded experiment testing two means of reducing aggression in humans. One arm involved administering cyproterone acetate which renders recipients docile and impotent, the other arm of the study involved psychosurgical destruction of specific areas of the brain believed to be associated with violent uncontrolled actions. Doe and his parents agreed to Doe's assignment to the psychosurgery treatment group, but the research came to a halt before any invasive procedures had begun because of a suit initiated by Michigan Legal Services attorney Gabe Kaimowitz. Doe initially resented the intrusion and was interested in continuing the experiment, however another (independent) legal appeal resulted in his release from prison. Subsequently, Doe withdrew his agreement to participate in the experiment. In spite of Doe's release, the Wayne County Circuit court believed that the issues raised by the case were not moot and they ultimately ruled in part that patients involuntarily committed in state institutions are legally incapable of giving competent, voluntary, knowledgeable consent to experimental psychosurgical operations:

it is impossible for an involuntarily detained mental patient to be free of ulterior forms of restraint or coercion when his very release from the institution may depend on his cooperating with the institutional authorities and giving consent to experimental surgery [29].

Case II: Lane v. Candura

Mrs. Candura was a 77-year-old diabetic widow who developed gangrene in her right foot and lower leg. she had had two earlier amputations (involving a toe and a part of her right foot). Her physicians recommended that the leg be amputated without delay, but after some vacillation, she refused the operation and persisted in her refusal. The trial court found that Mrs. Candura was:

incapable of making a rational and competent choice to undergo or reject the proposed surgery to her right leg. To this extent her behavior is irrational. She has closed her mind to the extent that the court cannot conclude that her decision to reject further treatment is rational and informed . . . [31].

The Court of Appeals, however, reversed this decision, writing that:

Mrs. Candura's decision may be regarded as most unfortunate but on the record of the case it is *not* the *uninformed decision* of a person incapable of appreciating the nature and consequences of her act ([31], p. 1236).

The court found that the competency underlying the decision was not to be determined by what Mrs. Candura's *physicians* believed to be rational, but rather that she understood the nature of the "proposed operation" and the consequences of refusing it. The court noted that Mrs. Candura "has made it clear that she does not wish to have the operation even though the decision will in all likelihood lead shortly to her death" ([31], p. 1235).

Case III: Roe III

The next case I want to review is known as the *Roe III* case decided in 1981 in Massachusetts.[11]

Richard Roe, III (a pseudonym) was committed to Northampton State Hospital in 1980 as a result of being charged with attempted unarmed robbery and assault and battery. Roe had been committed in the previous year and had been diagnosed as having schizophrenia, undifferentiated type. The diagnosis was reconfirmed on his second admission with the type now changed to paranoid schizophrenia. While in the institution he attacked another patient, and also refused to be treated with antipsychotic medication or to engage in any psychotherapy. Roe's refusal was claimed to have been based on his previous experience with illicit drugs that had contributed to his involvement in an automobile accident. It was alleged that Roe now accepted certain tenets of the Christian Science faith, though the judge concluded differently. Roe's father sought to be appointed his guardian and the issue of how and on what grounds Roe might continue to refuse treatment was brought before the Massachusetts Supreme Court. This court ruled in 1981 that the *guardian* could *not* authorize antipsychotic medical treatment, and that *only a judge* could decide on whether medication could be given to Roe III. The judge was to employ the "substituted judgment" standard developed in the well-known *Saikewicz* case. Citing the *Saikewicz* precedent, the court determined that the following "factors" must be taken into account in this type of case

in order to identify the choice which would be made by the incompetent person, if that person were competent, but taking into account the present and future incompetency of the individual as one of the factors that would necessarily enter into the decision-making process of the competent person.

These factors were given as follows:

(1) the ward's expressed preferences regarding treatment, (2) his religious beliefs, (3) the impact on the ward's family, (4) the probability of adverse side effects, (5) the consequences if treatment is refused, and (6) the prognosis with treatment.

Though Roe III's global *competency* is not at issue in the deliberations here, what is being decided is whether to allow a guardian to override Roe III's expressed preferences not to receive psychoactive drug treatment.

Case IV: ECT

The last case for consideration is a *Hastings Center Report Case* Study on "Saying 'No' to Electroshock" [47].

Mr. Clarence Arnold, a frail looking 60-year-old, was brought to the hospital by his wife following threats by him to shoot her and himself. He had a two-year history of weakness and numbness in his extremities, as well as headaches and dizziness. He also expressed fears of dying from "brain cancer". For the six months prior to admission, he reported trouble sleeping, weight loss, and admitted to anxiety attacks and delusions. An extensive physical examination was within normal limits. Mr. Arnold could not tolerate antidepressants and he refused to eat. Electroconvulsive therapy (ECT) was recommended but, after initially agreeing to it, he declined to begin treatment. A significant effort with the aid of two additional psychiatrists was made to persuade him to reconsider, which he did. After six treatments, however, he refused to continue and repeated efforts to get him to change his mind were unsuccessful. A second trial of antidepressants also failed. Eventually, Mr. Arnold agreed to a continuation of the ECT, but then refused to follow instructions. His condition continued to deteriorate. He refused to consider a *third* trial of ECT and was told, with his family's agreement, *that if he did not acquiesce, he would be committed and treated with ECT anyway.* The ECT was administered, and the patient had a favorable response, and is still doing well outside the hospital [47].

IV. MULTIAXIAL MODEL OF THE COMPETENCY CONCEPT

I now wish to begin to develop what I view as a more complex model of competency than any thus far presented — one which I think may be more adequate to the richness and objective complexity of competency determinations, whether these be legal or the more frequent medical or psychiatric decisions (which may precede a legal determination).

The thesis that I want to explore is that there are at least three interacting types of considerations involved in a competency determination. The *first* and easiest to defend can be termed the consideration of *rationality* or of *understanding*. This is the element which is common to most of the literature on competency surveyed above. It is the area of focus of Roth *et al.* in both their accounts of increasingly stringent tests and increasingly stringent capacities.

The Rationality/Understanding Axis

This is the component which psychiatrists are at present most comfortable assessing. Gutheil and Appelbaum's *Handbook of Psychiatry and the Law* ([23], pp. 217—220) offers specific suggestions for evaluations in this area and Weithorn [52] has developed a specific prototype instrument for operationalizing Roth *et al.*'s approach to competency assessment. Following Gutheil and Appelbaum, for general competency a patient should have:

> (1) an awareness of the nature of his situation, including his level of impairment if any exists; (2) a factual understanding of the issue or issues with which he must deal; (3) the ability to manipulate information rationally in order to reach a decision of these issues ([23], p. 217).

The third category is assessed with the assistance of a standard mental status examination testing for "orientation, intellectual functioning, judgment, impairment in rationality (hallucinations and delusions), and alterations of mood" as well as additional cognitive and affective dimensions ([23], p. 218). Any deficiencies require further work-up using specific examples, such as hypothetical business transactions ([23], p. 218).

Gutheil and Appelbaum also propose a set of tests for *specific* competencies, such as competency to consent to psychiatric treatment using psychotropic drugs. In such cases, the above three categories are particularized to the patient's psychiatric problem and the risks, benefits, and possible alternatives to the proposed treatment regimen. A patient must also have had an appropriate opportunity to learn the information necessary to provide an effective informed consent.

The Aggregate Expected Benefits/Harms Axis

The *second* component of the model I favor adopts a variant of Drane and others' sliding scale approach. As noted earlier, most commentators feel that a competency determination needs to take into account the risks and expected benefits to the patient of a finding of competency or incompetency. I view this component as an *initially* independent *dimension* of a competency assessment, but one then to be factored into the final competency assessment as I will suggest further below.

This second dimension can perhaps best be described as a generalization of the risk/benefit aspect of a competency determination. I think it might best be called the aggregate expected benefit/harms dimension, since it will comprise both risk/benefit considerations for the patient *as well as for others*. Thus a patient's dangerousness to others would be considered in this category. This aggregate determination is, I believe, composed in part by those complex factors suggested by the *Roe III* decision quoted above, including: a ward's expressed preferences, the impact on the ward's family, the probability of side effects, and the prognosis under both treatment and non-treatment options.[12]

Though it may initially appear odd that the issue of aggregated harms and benefits be treated as an aspect of *competency*, a notion whose root meaning as Culver and Gert have suggested is one of an 'expected ability', what we are discussing is the broader concept of a competency *determination*. This involves the idea of "competent enough in respect of capacity x to make decision y and to act (or to have others act) on that decision". Also, as noted in the case summary above on pp. 260–261, though the *Roe III* discussion does not question the general incompetency of the patient, the factors that the court indicated need be considered in determining whether to honor Roe III's preference not to be treated involve just these aggregated harm/benefit issues. I interpret this as tantamount to a judgment of a specific competency though one that could be overridden, but only by the court and not by the guardian.

The Volitional Axis

Following Cassell, Siegler, and Sherlock, I want to add still another consideration to a competency determination. This type of consideration has been termed by them the 'will' or *volitional* component. It is intended to capture those factors which adversely affect free choice but which do not appear to be purely conceptual or 'rational'. Siegler cites the interference with the patient's "true will" which "a reaction to the pain, fear, and uncertainty of his critical condition" can produce. Sherlock gives as examples command hallucinations and phobias, but he also cites the indecisiveness and unconcern frequently seen in depressed patients.

Other commentators have noted the importance of 'affective' features in clinical situations and their consequence for competency. Appelbaum and Roth wrote:

If the treatment or procedure for which consent is being sought is sufficiently provocative of anxiety or fear, the patient may be forced to revert to more primitive, even psychotic, levels of defense for coping with it ([2], p. 1463).

After presenting two cases exemplifying this type of clinical influence on a patient's choice, they concluded by saying: "An assessment of the psychodynamic basis of the patient's refusal should be part of every competency examination" ([2], p. 1464).

Appelbaum and Roth also note other clinical situational and interpersonal factors such as the "effect of the setting" (e.g., the races or the social classes of the patient and physician) which may affect "the patient's ability to formulate a competent decision" ([2], p. 1466).

In his correspondence with Appelbaum referred to above in note 11, Professor George Dix also addressed this issue and wrote:

... doesn't competency, properly conceptualized, involve an *ability* to resist certain influences? If so, it might be possible to regard a subject as incompetent because of the subject's established inability to resist certain types of influences and thus to avoid a perhaps more difficult inquiry into whether the subject did *in fact* remain unaffected by them, i.e., whether the subject made a "voluntary" decision.

If competency involves the ability to resist certain influences as well as the ability to understand (and perhaps use in a cognitive sense) certain factual information, perhaps the discussion of the *Kaimowitz* court is not off the mark (July 2, 1981, p. 2).

I should add, however, that the situation is not quite as clear as Professor Dix suggests, since *Kaimowitz* has been criticized for conflating a requirement for freedom of choice with an assessment of mental capacity ([3], p. 952).

Is the Triaxial Concept of Competency Too Broad?

Before returning to a more in-depth discussion of the volitional axis, which I believe needs more explication and defense than either of the other two axes of the model, it would be wise to pause for a moment and consider a recurrent objection which this analysis has encountered.[13] Meisel has put the objection in the following terms:

> ... competency is a *quality* of a person; we will not honor a person's decision if he lacks that quality. By contrast harm is not a quality of a person, but a *consequence* of an action or, as in the current context, a decision. Finally, voluntariness is partially a quality of a person, but it is also affected by a variety of *external forces* that come to bear on persons. A person is competent to make a decision regardless of the consequences of that decision or the internal or external forces that come to bear on him in the process of decisionmaking. Competency is a necessary, but not a sufficient, condition for an individual to have his decision honored.[14]

I characterize Meisel's account as expressed in the above quotation as the 'narrow' interpretation of the notion of competency. Meisel is quite willing to allow factors representing the harm/benefit ratio of the consequences and the absence or presence of volition-compromising conditions to function in a decision about whether or not a person's decision should be honored, but balks at "burdening the concept of competency" by incorporating into it the other two factors.[15]

I see the situation differently. In my view, an individual's *ability* to understand and appreciate the seriousness of the consequences of his decision for himself and for others is a *capacity* which is implemented in the situation which gives rise to the question of competency. Moreover, serious adverse consequences will not, in the account I am urging, *necessarily* result in the person's decision not being honored;[16] rather the individual must, to be allowed to act on such a decision, exhibit a correspondingly high degree of understanding and volitional autonomy.[17]

A similar set of considerations comes into play as regards the volitional component of competency. Here it is the *capacity* of the person to resist adequately either internal or external pressures affecting his decision-making ability. This *ability* is an important part of the individual's personality, which if it is compromised in the context of leading to potentially serious harms, can be interpreted as an integral aspect of that person's 'competency'.

From my perspective, *incorporating but not conflating* the aggregate expected harms/benefits and volitional axes *explicitly* into a model of competency offers the opportunity for recognition of these elements which often do affect competency determinations. Such explicit incorporation can also underscore the need for further clarification of those difficult-to-assess dimensions such as aggregation of expected harms and benefits and voluntariness.

V. CHARACTERIZING AND INTERPRETING THE VOLITIONAL AXIS

Rationale and Possible Approaches to Characterization and Interpretation

The inclusion of what I have termed a volitional component to competency is controversial, and other commentators have felt this is better analyzed under another rubric. They might propose to consider it either under the 'understanding' aspect of a competency determination — perhaps in terms of an 'affective' interference with rationality or as a failure in 'appreciation' of the information. The President's Commission suggested that 'free' volition or 'voluntariness' be treated as an independent requirement underlying effective patient participation in health care decision-making. It will thus be useful to spend some time discussing whether this volitional component of a competency is characterizable and assessable. In addition, the introduction of the issue of volition raises the specter of the typically interminable 'free will-determinism' debate and requires some means, even if only in outline form, for quarantining that red-herring. I will first briefly review Gert and Duggan's work and then turn to a somewhat different source, Hart's writings on voluntary activity and the types of excuses that are legitimately offered for non-voluntary conduct, to provide a *context* in which the free will-determinism issue can be (at least *temporarily*) laid to rest. I shall also briefly comment on current psychiatric sources of possible help in elucidating the volitional area.

Philosophical Approaches

Probably the first systematic treatment of voluntary, involuntary, and un- or non-voluntary conduct can be found in Aristotle's *Michomachean Ethics* (Book, III, i) [5]. For our purposes, however, I will begin by discussing the work of Gert and Duggan [16], Culver and Gert [10], and Gert writing with Ferrell *et al.* [19] who have elaborated a notion of *unvoluntary but intentional acts*. This category of acts is one that has been generally overlooked by both philosophers and jurisprudents. They have also introduced the notion of a "volitional disability" to refer to a *kind* of activity that certain people are unable to perform. They give as an instance the inability of a claustrophobic to enter an elevator, but believe that the category appropriately covers a number of mental illnesses including phobic, compulsive, and addictive disorders. Culver and Gert provide a rigorous definition of a "volitional ability" in terms of which a volitional *dis*ability can then be expressed. The definition of a volitional ability is an operational one, in part, and involves the probability near one that an individual will *will* (*or not will*) to perform the act if there are coercive incentives for doing (or not doing) so, and that there is a probability greater than zero that an individual will *will* (*or not will*) to perform the act if there are *non*coercive incentives for doing (or not doing) so [10].[18] The basic idea behind this complexity is the ability of an individual to *will*, given appropriate incentives. Even given *coercive incentives*, a claustrophobic *cannot will* to enter an elevator.

In their recent book on *A History and Theory of Informed Consent*, Faden and Beauchamp propose three conditions of "autonomous actions": (1) Intentionality, (2) Understanding, and (3) Noncontrol ([18], Ch. 7). They note that the third condition is typically defined in a "negative" manner using as basic concepts *influence* and *resistance to influence*, and they add that "these problems of control are often addressed in the literature on informed consent through the concept of voluntariness . . ." ([18], p. 256). Faden and Beauchamp view influences as lying on a continuum, and provide a suggestive diagram which represents the spectrum of such influences. This is reproduced below as Figure 1.

I believe that the elaboration of the notion of controlling influences, such as has been done by Faden and Beauchamp, and development of concepts of volitional ability and volitional disability by Gert, Culver, and their colleagues, are useful though still preliminary ideas. These notions can perhaps both be broadened and given additional content so as to refer to a wider variety of *kinds of situations* in which specific individuals in specific time-frames may not possess the ability to *resist*

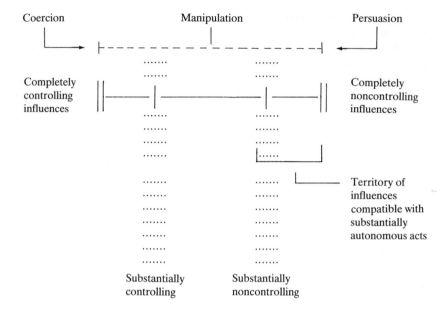

Figure 1. The continuum of influences from controlling to noncontrolling. (After Faden and Beauchamp, [18] p. 259.)

certain influences such as fear of a diagnostic or therapeutic procedure.[19] Under this view, an individual may have a proper understanding of a diagnostic/therapeutic procedure, in terms of being able to cite the nature and consequences of it, but may not be able to agree to the procedure because of a *volitional* disability. Such an individual may even be able to offer pretexts as to why he is refusing the procedure, which may or may not be credible to a physician or a judge. (I am sensitive to the potential abuse to which such a notion could be put, in the sense that it could be used to justify stronger forms of paternalism. I also think, however, that proper safeguards could be established to prevent such abuses, such as public and explicit criteria for volitional disabilities.)[20]

A Jurisprudential Context for Volition: Hart's Approach

As noted above, the introduction of the volition issue into a competency determination raises the problem of the often inconclusive free will-determinism debate. We need to have a sense of the voluntary and the involuntary (and unvoluntary) act which is located within the legal and *practical* ethical realm. It is my sense that we can obtain the necessary perspective by building on some of the suggestions made by the eminent jurisprudent H. L. A. Hart.[21]

Hart, in a penetrating 1958 paper [25] on 'Legal Responsibility and Excuses', explored the nature and ground of *excusing conditions* in the criminal law and their civil analogues known as *invalidating conditions* covering such civil transactions as wills, gifts, contracts, and marriages. He wrote that:

If an individual breaks the law when none of the excusing conditions are present, he is ordinarily said to have acted of "his own free will", of his own accord, "voluntarily"; or it might be said "He could have helped doing what he did" ([25], p. 95).

This approach implicitly suggests that one means of characterizing voluntary behavior might be to *exclude* the reasonable possibility that any of these excusing/invalidating conditions apply. In point of fact, this approach suggests a rather *general* means of describing and in principle testing for those situations in which a diminished competence due to diminished voluntary activity is to be found.

In his article, Hart examines two proffered explanations for the existence of such excusing/invalidating conditions, which he rejects, proposing his own account of their function. To Hart, such conditions maximize "the efficacy of an individual's informed and considered choice in determining the future and also his power to predict the future" ([25], p. 111). From the perspective of such a function, civil transactions are appropriately invalidated by the "mental factors" of "mistake, ignorance of the nature of the transaction, coercion, undue influence, or insanity". To Hart this is so because:

... a transaction entered into under such conditions will not represent a *real choice*: [emphasis added] the individual might have chosen one course of events and by the transactions procured another (cases of mistake, ignorance, etc.), or he might have chosen to enter the transaction without coolly and calmly thinking out what he wanted to do (undue influence), or he might have been subjected to the threats of another who had imposed *his* choices (coercion) ([25], p. 110).

Hart's thesis that real choice requires the absence of the excusing/ invalidating conditions, coupled with Gert, Culver, and Duggan's proposal of the existence of a class of volitional disabilities suggests three conclusions.

(1) A volitional disability may be appealed to in a competency determination but it will have to be appropriately both narrowed and further elaborated. It will have to be properly narrowed in the sense that the disability be restricted to those situations in which competency determinations arise and where a consensus emerges that *this kind* of volitional disability (to this degree) is analogous to Hart's excusing/ invalidating conditions. (Thus the determination will also have to be individualized to the patient in question.) Suitably elaborated to cover an appropriate range of psychiatric disorders, a volitional disability may be tested for and perhaps even quantitated. A necessary caveat that should be added is that such a disability may fluctuate under different environmental conditions and over time, a situation analogous to that stressed by such commentators on competency determinations as Appelbaum and Roth [4], Murphie [37, 38], and Beauchamp and McCullough [7].

(2) The presence in an individual of a serious voluntary disability which after assessment is judged as reasonably constituting an excusing/ invalidating condition is prima facie evidence of a correlative (but possibly *quite restricted and specific*) *incompetency* of that individual.

(3) Just as the function of the excusing/invalidating conditions is maximally to protect the individual in his or her free choice, so a determination of incompetency is the corresponding practice whose function is to protect an individual against an involuntary or unvoluntary choice.

Psychiatric Approaches

The standard practice of requesting psychiatric consultations in connection with competency determinations is warranted by the need to exclude 'insanity' in the patient by conducting a thorough mental status examination. But the psychiatrist may also have another more subtle role to play if the thesis that there is a volitional component to competency determinations is correct. A psychiatrist is presumably skilled at eliciting those determinants of behavior that are 'unconscious' but nonetheless directive of a patient's actions. A psychiatrist need not subscribe to Freudian doctrines to appreciate the important role that unconscious motivation may play in patient decision-making.[23] Jay Katz in his book, *The Silent World of the Doctor and Patient*, has underscored the significance of unconscious as well as other psychoanalytic factors and their effect on patient autonomy and the doctor-patient relation ([30], esp. chs. 5—6).

One difficulty with the use of current psychiatric approaches to assessing where a patient may fall on the volitional axis is the tendency to eschew etiological analysis in DSM-III.[23] In point of fact, a review of DSM-III, though it discloses a number of disorders which involve compromised volition, such as the Obsessive-Compulsive disorder and the 'Pathological Gambling' diagnosis (in which there is a "failure to resist impulses to gamble"), has essentially nothing useful to say concerning the testing and assessment of explicitly volitional disabilities, nor does DSM-III-R address this lacuna.

The prospect of developing appropriate instruments and measures to assist in the characterization of the volitional axis is not necessarily bleak, however. Various legal instruments for the assessment of competence have been investigated in recent years, and the employment of structured patient interview techniques supplemented with specific scenarios about which the patient is asked to comment shows some promise of revealing underlying factors affecting volition.[24]

I am going to close this section by anticipating a possible objection that might be raised against incorporating a volitional axis into a competency assessment, namely, that the recognition of any such axis would downgrade the role of an individual's responsibility for his actions.[25] Thus a drug addict might allege that he has a volitional disability which prevents him from breaking his drug habit. He might also argue that he has a correlative volitional disability which compels him to steal to support that habit.

The proper response to such an objection is to recall that we are speaking of a notion of volitional disability characterized by *degrees*, not of a binary or an all-or-none concept. In my view, it is the role of the social system through its legislative bodies and courts to define via its laws and precedents where to draw the line between culpable and non-culpable actions. The social system does so, referring again to Hart's views, by appealing to specific kinds of volitional situations which are exculpatory, through its use of excusing/invalidating conditions. A society may well draw that typically non-bright line between the culpable and the non-culpable at different points as it experiments with attempts at efficient and fair reforms in its legal system; a similar reassessment can be carried out in the realm of moral justification as well.

Thus the admission of a volitional axis into a competency determination does not automatically eliminate an important role or determinate standards for individual responsibility for one's actions. Law (or ethics) may *refuse to acknowledge* a volitional disability of some specified strength in a specific set of circumstances as an excusing condition, and be able to provide reasons for such a refusal.

VI. WHAT MIGHT A COMPETENCY MODEL WITH A VOLITIONAL AXIS ADD?

I now want to look both at the cases introduced above as well as some additional bioethics cases in the light of the multiaxial model for competency determination.

The Four Test Cases

(1) *Kaimowitz*. If we examine Louis Smith's situation with respect to the triaxial model proposed above, we encounter some interesting disparities along the three axes. On the basis of the evidence, including the detailed consent form which spelled out the risks and expected benefits of the psychosurgery procedure, there is little to question in the cognitive dimension. As regards the volitional, however, the situation is less clear. Annas and his colleagues have commented on this case citing Erving Goffman's account of the effects of total institutions, noting that such institutional settings make it "difficult for one not to feel some sort of coercion or encouragement to consent merely in being approached for the particular procedure" ([1], p. 3–15). Finally, with respect to the degree of risk involved to the patient and the benefits expected to accrue to the patient and to society, the research program was viewed by the court as profoundly uncertain, holding that "the lack of knowledge on the subject [of psychosurgery] makes a knowledgeable consent to psychosurgery literally impossible" [29]. Though the court's decision is widely viewed as turning on the issue of diminished volition, and has been seen by Annas *et al.* as therefore setting an inappropriate precedent, I believe that an alternative reading is more likely. Under this alternative, the compromised volition is only a contributing factor in the assessment, with the risks of an invasive procedure being at least as strong a deterrent to court approval. Thus I interpret the decision in terms of the interaction of the three axes discussed in the model as a sound one.

(2) *Lane v. Candura.* In this case, attention to the possibility of a diminished volition suggests that additional information would have been warranted. Some further explanation of the reasons, with special attention to the *authenticity* of Mrs. Candura's preferences, would have been in order. In this type of situation, the risks (of death) are high, but may well be counterbalanced by the benefit of a 'good death' depending on the patient's situation. A careful review of the factors suggested by Siegler and Goldblatt would have been in order. Robertson does note, citing Candura as well as two other similar cases, that "a consensus has emerged that competency to make medical decisions may coexist with civil commitment, fear of the consequences, intermittent lucidity or disorientation, and even irrational behavior" ([4], p. 562). Though there was a court sanction in this case (as Robertson adds at a later point in his essay), in similar but unsanctioned circumstances one can run the risk of abandonment of a patient that may "lead to civil and even criminal liability in extreme cases" ([41], p. 568). From my perspective the *Candura* case is a 'hard case', one that suggests we need both further data as well as better means of assessing that data for volitional compromise.

(3) *Roe III.* In my view, this case turns on the aggregated risk/benefit considerations. There is a diagnosed thought disorder present, but no clear evidence that the patient's preferences, understanding, or volition are seriously compromised in the specific area of refusal of psychoactive drug treatment. There are also serious side effects to antipsychotic drugs, counterbalanced to an extent by the benefits to the patient and his society of control of his illness. The balancing of individual versus social benefits is a complex and difficult task, but not one that is specific to this case. I see the competency model in such situations as helping to identify the areas in which additional information and/or social theories will be useful.

(4) *ECT.* Any comment on this case needs to be sensitive to the fact that Mr. Arnold represents that unfortunate type of case in which a *good outcome* may prejudice us against judging a *bad procedure* as bad. As with Mrs. Candura, and in the light of Mr. Arnold's severe depression, more information about the reasons for his refusal would have been helpful. Mr. Arnold is a patient who indicates the need to develop those assessment tools which would help elucidate the volitional dynamics of his refusal. There does not appear to be any question about Mr. Arnold's *understanding* as regards ECT, and the

procedure is, for all its poor press, not a dangerous one. Since no alternative therapeutic options remained available, and because of the risk to the patient of suicide, the 'pain' of depression, and the benefits of remission to Mr. Arnold's wife and to society, the aggregated expected harms/benefits judgment inclines strongly toward treatment. In the light of a highly probable volitional disability, Mr. Arnold's treatment was, I think, indicated despite his refusal.

Other Bioethics Cases

There are other cases in the biomedical ethics literature on autonomy and paternalism that provide evidence for the utility of the triaxial model outlined above. Jackson and Younger in their article on 'Patient Autonomy and "Death with Dignity"' [28] provided descriptions of six patients which exemplified "clinical and psychological problems that may complicate the concept of patient autonomy and the right to die with dignity" ([28], p. 405). Jackson and Younger's patients' decisions are *influenced* by such psychodynamic factors as ambivalence, depression, hidden problems, and fears. Their patients were not able to resist the effects which their illness and social situation exerted on their decision-making capacity. Cassell's cases in his essay on 'The Function of Medicine' [9] also suggest that the triaxial approach has merit, and Siegler's patient, Mr. D [49] in the light of siegler's later comments [50] and in view of Katz's comments on the case [30], underscored the need to take into account not only a patient's understanding of the nature and consequences of refusing treatment, but also the status of that patient's volitional freedom, all within the context of the joint expected harms and benefits.

VII. SUMMARY AND CONCLUSION

In this paper, I have argued for a type of 'sliding scale' model of competency, but one that is considerably more complex than Drane's. One can, though it is not necessary to the model's application, represent the model pictorially (see Figure 2) as a set of three mutually perpendicular axes involving (1) rationality or understanding, (2) volitional freedom, and (3) aggregated expected benefits/harms.

There are constraints on permissible ranges in such a space which require a combination (perhaps with a specification of lower-limit

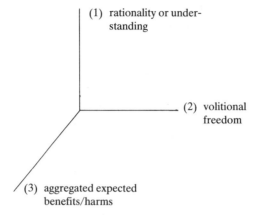

Figure 2. Pictorial Representation of the Triaxial Model.

thresholds) of understanding and volitional freedom for situations with decreasing expected benefits. Thus a low risk/high gain procedure can be refused by a patient only if he is located in the high understanding and high volitional freedom region. Some means of combining considerations, and illustrating a range of types of cases, is needed to make this notion clearer, and in the previous section I have attempted to begin this task. We will not, however, even if more extensive formal instruments for assessment are developed, be able to find an *algorithm* for dealing with competency cases. As Grisso suggests, such highly developed assessment instruments suitable for introduction into the legal context

may provide standardized information that can clarify the extent of a patient's relevant capacities, and they can promote a consistent quality of information across competency cases. Yet no particular assessment result will relieve the professional of the responsibility for discretion in dealing with the moral quality of the judgment ([22], p. 325).

It might be useful in this closing section to offer a brief comparison of the above model with a comprehensive account recommended by Beauchamp and McCullough in their monograph *Medical Ethics: The Moral Responsibilities of Physicians* [7]. This analysis, as noted earlier, draws on the contributions of a number of research programs in competency determinations, synthesizes them, and offers a useful perspective for several further comments on model application.

Beauchamp and McCullough state that "All judgments of competence require the specification of a context, of relevant abilities, of the stability of the abilities at a time, and of some threshold degree of possession of the abilities" ([7], p. 118). Context is required because, as I have noted above, a person may be competent to do one set of tasks yet incompetent for another. In general, the particular problem generating the question of competency will determine the context, but attention to *specific* competence determinations, as indicated in the discussion of Gutheil's and Appelbaum's suggestions earlier, is appropriate. I view specification of the context, as well as the relevant abilities under question, as part of the *application instructions* for the model I proposed above.

Beauchamp and McCullough's suggestion that we pay careful attention to the stability and variability of the abilities is, in my view, quite correct, though I view this as a rather *general* caveat for *all aspects of patient management*. Since there may be a tendency to overlook patient variability with respect to both intellectual comprehension as well as volitional freedom, it is appropriate to note the possibility, but I do not see such variability as part of a model of competency.

Finally, I want to agree with Beauchamp and McCullough's proposal that "competence is a continuous concept", rather than a simple threshold notion,[26] but I want to argue that Beauchamp and McCullough have not identified the fact that the concept of competency requires an examination of this continuity along three prima facie independent (though ultimately related) dimensions of understanding, volition, and expected harms/benefits. Thus I see the model offered above as a superset of the Beauchamp and McCullough proposal, and one that both is more congruent with the complexities of professional decisions in competency cases, as well as one that suggests where additional investigation is needed to resolve still very difficult and imporotant issues.[27]

George Washington University
Washington, D.C., and
University of Pittsburgh
Pittsburgh, Pennsylvania

NOTES

[1] A tendency has emerged recently, perhaps generated initially by the President's Commission analysis of competency [39], to distinguish the concepts of 'competency'

and 'capacity' (in the sense of 'decision-making capacity'). This distinction can be found both in recent statute law (in the 1988 New York State Law regarding foregoing life sustaining treatments) and in ethical discussion (see the *Hastings Center Guidelines* [26], esp. pp. 131—133). Though there is some rationale in the distinction — as noted below a competency determination is a *legal* decision, whereas an assessment of decision-making capacity can be done by both physicians and layfolk — much of the discussion in the ethical literature utilizes the vocabulary of 'competency' and covers both legal and ethical issues in terms of that language. Thus, 'competency' terminology will be used in the present paper, with specific indications where there may be some unclarity as to whether the ethical or the legal sense is meant. (Interestingly, Buchanan and Brock's recent book, *Deciding for Others: The Ethics of Surrogate Decision Making* (New York: Cambridge University Press, 1989), uses the term 'competence' extensively, even though the authors construe the notion as referring to "decision-making capacity" (p. 18).)

[2] For a discussion and analysis of the fundamental bioethical principles of autonomy and beneficence, see The Belmont Report [8], Beauchamp and Childress [6], and President's Commission volume on *Making Health Care Decisions* [39], esp. pp. 41—51.

[3] See for example M. Siegler's case of Mr. D. [49]. Appelbaum and Roth [4] provide data indicating that only 7 of 105 cases of treatment refusal involved psychiatric consultation.

[4] For a celebrated example, see the case of Dax Cowart in [52]; also see [7], pp. 79—82.

[5] For an introduction to the issue of rational suicide see [33].

[6] See Meisel and Roth [34] and Beauchamp and McCullough ([7], pp. 123—124). Beauchamp and McCullough's classification of Meisel and Roth's seven tests involves placing them under the following four categories: (1) absence of a decision, (2) absence of an adequate reason (in making a decision), (3) absence of an adequate understanding (in making a decision), and (4) absence of an adequate outcome (in making a decision).

[7] See R. Reiser's discussion of limited guardianship in his [40], pp. 491—492.

[8] M. Siegler and A. D. Goldbatt, ([50], p. 13). This essay is a development of an earlier article by Siegler [49]. For interesting comments by Jay Katz on the patient discussed by Siegler in his [49], see Katz ([30], pp. 159—163). But also see Siegler in [48], Ch. 1.

[9] Sherlock's position was anticipated by George Dix in correspondence with Appelbaum in a letter dated July 2, 1981 [11] which commented on the *Kaimowitz* case [29].

[10] See, for example Roth, Meisel, and Lidz who wrote: "in practice [competency] . . . seems to be dependent on the interplay of two other variables, the risk/benefit ratio of treatment and the valence of the patient's decision, i.e., whether he or she consents to or refuses treatment" ([44], pp. 282—283). Also see B. Miller or four increasingly strict senses of autonomy [37], Cassell [9], and Siegler in [48], p. 59.

[11] See [42]. For selections, see Reiser [49], pp. 509—516. Also see Gutheil and Appelbaum [24] for an interesting criticism of the substituted judgment doctrine.

[12] There are formidable problems involved in developing a precise characterization of aggregation of expected benefits and harms. This is a complex problem in social choice theory, welfare economies, and theoretical utilitarian analysis, all of which can shed useful light on the issues. Suffice it for this essay to say that psychiatrists and judges do in practice perform such aggregations, and have, as noted, provided rather general accounts of the factors which should be considered.

[13] The objection to be discussed below has come principally from Professor Alan Meisel and Dr. Loren Roth in comments on earlier versions of this essay.

[14] A. Meisel, personal communication, 10-13-86.

[15] A. Meisel, personal communication, 10-13-86.

[16] The health care system (and society's) willingness to permit serious harmful behavior will of course differ between purely self-regarding behavior and behavior which adversely affects other persons.

[17] Miesel's concerns may reflect a Janus-faced feature of a competency determination. In such, there is the need for a *patient* to appreciate the situation, in the sense of appropriately weighing the burdens and benefits of an action (or non-action); but there is also the need for the *physician/judge* to evaluate the seriousness of the burdens and benefits of the patient's action/non-action in the light of the patient's preference and capacity for assessment/appreciation.

[18] Culver and Gert's terminology does not involve probability but used language such as almost always, sometimes, and the like. I view this as equivalent to the probabilistic formulation.

[19] Meisel, Lidz, and Roth have discussed the notion of a catalogue of different types of cicrumstances in which pressures are brought to bear on individual's decision-making which may attenuate voluntariness. See Meisel and Roth [35], esp. pp. 309–315, and Lidz [32], esp. pp. 119–122.

[20] See footnote 24 below regarding the possibility of developing such explicit criteria in the form of legal instruments possessing reasonable levels of validity and reliability.

[21] What is offered here based on Hart's approach is a 'practical' means of bracketing the free will-determinism problem; it is not meant to be an in-depth analysis. I believe that a useful further step could be taken on this issue by expanding and applying in this type of context Frankfurt's [20] notion of second order desires, but to do so would take me beyond the scope of this paper.

[22] For a non-Freudian discussion of the importance of the unconscious, see the discussion by Strayhorn ([51], pp. 193–196).

[23] The recent *DSM-III-R* revision does not alter this approach.

[24] T. Grisso in his recent book [22] indicates that Shinn and Sales in an unpublished paper discuss several recommendations for developing assessment instruments to measure patients' abilities to resist coercion ([22] p. 322). There is also an unpublished pilot study by Lippo *et al.* cited by Meisel and Roth ([35], p. 310, n. 259), which has studied different 'influence variables' effects on role-playing patients. The role of 'willing' and how to assess it empirically has been studied by Howard and his associates [27].

[25] This objection was first raised by C. Lidz in discussion.

[26] Continuity and threshold notions are not exclusive, and a threshold aspect could be incorporated into the model presented in the current paper. For more discussion on this issue, see Beauchamp and McCullough [7], p. 120, n. 24, and also Wikler's [54].

[27] This paper is a further development of an earlier draft version written for a Senior Elective in 'Law and Psychiatry' in the Department of Psychiatry of the School of Medicine, University of Pittsburgh. I am indebted to Drs. Geller, Lidz, Roth, Wettstein and Amchin, and to Prof. Meisel, for guidance and comments on an earlier draft. The errors, of course, are my own responsibility.

BIBLIOGRAPHY

1. Annas, G. J., Glantz, L. H. and Katz, B. F.: 1976 'Law of Informed Consent in Human Experimentation: Institutionalized Mentally Infirm', in the Appendix to *Research Involving Those Institutionalized as Mentally Infirm*, U.S. National Commission for the Protection of Human Subjects, DHEW Publication No. (OS) 78-0007.
2. Appelbaum P., and Roth, L.: 1981, 'Clinical Issues in the Assessment of Competency', *American Journal of Psychiatry* **138**, 1462–1467.
3. Appelbaum, P. and Roth, L.: 1982, 'Competency to Consent to Research: A Psychiatric Overview', *Archives of General Psychiatry* **39**, 951–958.
4. Appelbaum, P. and Roth, L: 1983, 'Patients who Refuse Treatment in Mental Hospitals', *Journal of the American Medical Association* **250**, 1296–1301.
5. Aristotle: *Nichomachean Ethics*, London, Penguin Books.
6. Beauchamp, T. and Childress, J.: 1985, *Principles of Biomedical Ethics*, 2nd edition, Oxford, New York.
7. Beauchamp, T. and McCullough, L.: 1984, *Medical Ethics: The Moral Responsibilities of Physicians*, Prentice Hall, Englewood-Cliffs, N.J.
8. *The Belmont Report: Ethical Principles and Guidelines for the Protection of Human Subjects of Research:* 1978, DHEW Publication No. (OS) 78-0012, Washington.
9. Cassell, E.: 1977, 'The Function of Medicine', *Hastings Center Report* **7** (December), 16–19.
10. Culver, C. and Gert, B.: 1982, *Philosophy in Medicine*, Oxford, New York.
11. Dix, G.: 1981, Letter to P. Appelbaum dated July 2, 1981.
12. Drane J. F.: 1984, 'Competency to Give an Informed Consent: A Model for Making Clinical Assessments', *Journal of the American Medical Association* **252**, 925–927.
13. Drane, J. F.: 1985, 'The Many Faces of Competency', *Hastings Center Report* **15** (April), 17–21.
14. DSM-III: 1980, *Diagnostic and Statistical Manual of Mental Disorders*, 3rd edition, American Psychiatric Association, Washington.
15. DSM-III-R: 1987, *Diagnostic and Statistical Manual of Mental Disorders-Revised*, 3rd edition, revised, American Psychiatric Association, Washington.
16. Duggan, T. and Gert, B.: 1967, 'Voluntary Abilities', *American Philosophical Quarterly* **4**, 127–135.
17. Dworkin, G.: 1976, 'Autonomy and Behavior Control', *Hastings Center Report* **6** (February), 23–28.
18. Faden, R. and Beauchamp, T.: 1986, *A History and Theory of Informed Consent*, Oxford, New York.
19. Ferrell, R. *et al.*: 1984, 'Volitional Disability and Physician Attitudes toward Noncompliance', *The Journal of Medicine and Philosophy* **9** (4), 333–351.
20. Frankfurt, H. G.: 1971, 'Freedom of the Will and the Concept of a Person', *Journal of Philosophy* **68** (1), 5–20.
21. Green, M. D.: 1941, 'Judicial Tests and Mental Incompetency', *Missouri Law Review* **6**, 141–165.

22. Grisso, T.: 1986, *Evaluating Competencies: Forensic Assessments and Instruments*, Plenum Press, New York.
23. Gutheil, T. and Appelbaum, P.: 1982, *Clinical Handbook of Psychiatry and the Law*, McGraw-Hill, New York.
24. Gutheil, T. and Appelbaum, P.: 1983, 'Substituted Judgment: Best Interests in Disguise', *Hastings Center Report* **15** (June), 8—11.
25. Hart, H. L. A.: 1958, 'Legal Responsibility and Excuses', in S. Hook (ed.), *Determinism and Freedom in the Age of Modern Science*, Collier, New York, pp. 95—116.
26. Hastings Center: 1987, *Guidelines on the Termination of Life-Sustaining Treatment and the Care of the Dying*, Hastings-on-Hudson, New York.
27. Howard, G. S. and Conway, C. G.: 1986, 'Can There be an Empirical Science of Volitional Action', *American Psychologist* **4** (11), 1241—1251.
28. Jackson, D. L. and Younger, S.: 1979, 'Patient Autonomy and "Death with Dignity" ', *New England Journal of Medicine* **301**, 404—408.
29. Kaimowitz v. Department of Mental Health, Civil No. 73-19434-AW (Cir. Ct. Wayne County, Mich., July 10, 1973).
30. Katz, J.: 1984, *The Silent World of the Doctor and Patient*, Free Press, New York.
31. Lane v. Candura, 6 Mass. App. Ct. 377 N.E. 2d 1233 (1978).
32. Lidz, C. et al., 1984: *Informed Consent: A Study of Decisionmaking in Psychiatry*, Guilford, New York.
33. Mayo, D. J.: 1986, 'The Concept of Rational Suicide', *The Journal Medicine and Philosophy* **11**, No. 2, 143—155.
34. Meisel, A. and Roth, L.: 1981, 'What We Do and Do Not Know About Informed Consent', *Journal of the American Medical Association* **246**, 2473—2477.
35. Meisel, A. and Roth, L.: 'Toward an Informed Discussion of Informed Consent', *Arizona Law Review* **25**, 265—346.
36. Miller, B.: 1981, 'Autonomy and the Refusal of Lifesaving Treatment', *Hastings Center Report* **11** (August), 22—28.
37. Murphy, J.: 1974, 'Incompetence and Paternalism', *Archiv für Rechts- und Sozialphilosophie* **50**, 465—486; reprinted in [38].
38. Murphy, J.: 1979, *Retribution, Justice and Therapy: Essays in the Philosophy of Law*, Dordrecht, Reidel.
39. President's Commission for the Study of Ethical Problems in Medicine and Biomedical and Behavioral Research: 1982, *Making Health Care Decisions* Vol. **1**, U.S. Government Printing Office, Washington.
40. Reiser, R.: 1985, *Law and the Mental Health System*, West Publishers, St. Paul.
41. Robertson, J.: 1991, *This volume*, 127—148. (Also appears as 'The Geography of Competency', *Social Research* **52**, No. 3 (1985). [Page refs. to *SR*.])
42. [Roe III] In the Matter of Guardianship of Richard Roe III, 421 N.E. 2d 40 (Mass, 1981).
43. Roth, L., Appelbaum, P., Sally, R. et al.: 1982, 'The Dilemma of Denial in the Assessment of Competency to Refuse Treatment', *American Journal of Psychiatry*, **139**, 910—913.
44. Roth, L., Meisel, A. and Lidz, C.: 1977, 'Tests of Competency to Consent to Treatment', *American Journal of Psychiatry*, 279—284.

45. [Saikewicz] Superintendent of Belchertown State School v. Saikewicz, 373 Mass. 728, 370 N.E. 2d 417 (1977).
46. Sherlock, R.: 1984, 'Competency to Consent to Medical Care: Toward a General View', *General Hospital Psychiatry* **6**, 71—76.
47. Sherlock, R. and Haykal, R.: 1982, 'Saying No to Electroshock', *Hastings Center Report* **12** (December), 18—19.
48. Siegler, M.: 1986, in Jonsen, A., Siegler, M., and Winslade, W. (eds.), *Clinical Ethics: A Practical Approach to Ethical Decisions in Medicine*, 2nd edition, Macmillan, New York.
49. Siegler, M.: 1977, 'Critical Illness: The Limits of Autonomy', *Hastings Center Report* **7** (October), 12—15.
50. Siegler, M. and Goldbatt, A. D.: 1981, 'Clinical Intuition: A Procedure for Balancing the Rights of Patients and the Responsibilities of Physicians', in S. F. Spicker, J. Healey and H. T. Engelhardt (eds.), *The Law-Medicine Relation: A Philosophical Exploration*, Reidel, Dordrecht, pp. 5—31.
51. Strayhorn, J. M.: 1982, *Foundations of Clinical Psychiatry*, Medical Year Book Publishers, Chicago.
52. Weithorn, L.: 1980, *Competency to Render Informed Treatment Decision*, unpublished Ph.D. Dissertation, University of Pittsburgh.
53. White, R. and Engelhardt, H. T.: 1975, 'A Demand to Die', *Hastings Center Report* **5** (June), 9—10.
54. Wikler, D.: 1979, 'Paternalism and the Mildly Retarded', *Philosophy and Public Affairs* **8** (Summer), 377—392.

NOTES ON CONTRIBUTORS

Virginia Abernethy, Ph.D., is Professor of Psychiatry (Anthropology), Vanderbilt University School of Medicine, Nashville, Tennessee.

Tom L. Beauchamp, Ph.D., is Professor of Philosophy and Senior Research Scholar, Kennedy Institute of Ethics, Georgetown University, Washington, D.C.

Eugene V. Boisaubin, M.D., is Associate Professor for Clinical Medicine and Ethics, Director General Internal Medicine Programs, Department of Medicine, Baylor College of Medicine, Houston, Texas.

Mary Ann Gardell Cutter, Ph.D., is Assistant Professor of Philosophy, University of Colorado at Colorado Springs, Colorado Springs, Colorado.

Edwin DuBose, Ph.D., is Associate for Theology, Ethics, and Clinical Practice, Park Ridge Center, Chicago, Illinois.

Susan Hankin Denise, J.D., M.P.H., is Instructor of Legal Research and Writing, Georgetown University Law Center.

Edmund L. Erde, Ph.D., is Professor, Department of Family Practice, School of Osteopathic Medicine, University of Medicine and Dentistry of New Jersey, Camden, New Jersey.

James Knight, M.D., is Professor Emeritus of Psychiatry, Louisiana State University School of Medicine, New Orleans, Louisiana.

E. Haavi Morreim, Ph.D., is Associate Professor, Department of Human Values and Ethics, College of Medicine, University of Tennessee, Memphis, Tennessee.

Edmund D. Pellegrino, M.D., is Director, Center for the Advanced Study of Ethics, and John Carroll Professor of Medicine and Medical Ethics, Georgetown University, Washington, D.C.

Marc Perl, M.B.B.S., is Physician Specialist, Stanford Medical Center, Stanford, California.

Edmund L. Pincoffs, is Professor Emeritus, Department of Philosophy, University of Texas at Austin, Austin, Texas.

John A. Robertson, J.D., is Thomas Watt Gregory Professor, University of Texas, School of Law, Austin, Texas.

Kenneth Schaffner, M.D., Ph.D., is University Professor of Medical Humanities, George Washington University, Washington, D.C., and University Professor, Department of History and Philosophy of Science, University of Pittsburgh, Pittsburgh, Pennsylvania.

Earl E. Shelp, Ph.D., is Executive Director, Foundation for Interfaith Research and Ministry, Houston, Texas.

Harold Y. Vanderpool, Ph.D., is Professor, Institute for the Medical Humanities, University of Texas Medical Branch, Galveston, Texas.

Stephen Wear, Ph.D., is Assistant Professor, Department of Philosophy and Department of Medicine, State University of New York at Buffalo, Amherst, New York.

Patricia White, J.D., is Visiting Professor, School of Law, University of Michigan, Ann Arbor, Michigan.

INDEX

Abernethy, V. 3—4, 211—226
Ackerman, T. F. 12
acutely ill patients 34—38
age 36
Annas, G. 17, 134—135, 141
Appelbaum, P. 10, 17, 131, 135, 179, 180, 254—255, 262—263, 264, 270
autonomy 32, 38—40, 59—64, 83—85, 95—100, 100—105, 127—128, 185—186, 199, 227—236, 239

Beauchamp, T. 31, 38—40, 49—77, 97, 102—104, 112, 197, 218, 239—240, 242, 244, 245, 267, 270, 275—276
beliefs 38—40, 122(n.4)
beneficence 39, 185—188, 240
benefits (see risk/benefit ratio)
Benn, S. I. 121
Bersoff, D. 74(n.8)
Boisaubin, E. 167—177, 179—182, 184, 188—191

capacity 33, 50, 140, 187—188, 213, 254—258
Cassell, E. 15, 247—248, 257—258, 264
Childress, J. 36
choice 17, 30, 36, 63, 65, 180, 255
competency (*see also* autonomy, capacity, denial, rational decision-making, understanding)
  and action 56—57
  approaches to evaluating 16—20
  case law 129, 132—133, 149—160
  consistency 22, 36, 99—100
  contract and 149—153
  criteria 49—51, 64—72

cultural antecedents 211—224
definition 30, 50, 131, 197—209
diminished 95—100
disagreements among physicians 7—15
dubious 108—116
economic antecedents 211—224
epistemic problems of 105—108
evanescent 55—56
factors in the evaluation of 20—22
family role in determining 15—16
freedom and 227—236
function 52—55
gatekeeping function of 52—55
general 56—59, 227
impaired 108—116
law and 108—119, 128—131, 149—160, 212—214
legal 68—71
meaning 49—51
medico-legal 69—72, 108—119
moral issues 41—44, 66—67, 79—89, 185—192
non-psychiatric determination of 4—6, 29—44, 167—176
partial 55—56
psychiatric determination of 3—4, 6—15, 16—24, 179, 248—251
psychological 67—69
sliding scale criteria for 31, 258
social practice of determining xii
specific 56—59
substantial 55—56
task performance 21
tests 64—72
a triaxial concept 253—276
Culver, C. 23—24, 31, 132, 135, 255—256, 263, 267
Cutter, M. A. ix—xix

285

INDEX

delusion 10, 151, 183
denial 13, 22—24, 181
Denise, S. 149—163
depression 20
Drane, J. xvi, 44, 257—258
DSM-III 205—206, 271
DuBose, E. 185—194
Dworkin, G. 15

Engelhardt, H. T. xviii
Erde, E. 237—252

Faden, R. 267
Fire Baptized Holiness Christian 200—201
foolishness 83
Freedman, B. 240—241, 243

gatekeeping function 52—55
Gert, B. 31, 135, 163, 167, 255—256
Goldman, A. 96
Green, M. 152
guardians 40—41

hallucinations 10
harm 263
Hart, H. L. A. 80, 268—270

information 21, 31, 57—65, 255, 256
informed consent 61—62, 93, 267—268
intention 30
interests 246
irrational decision-making 10, 81—83

Jackson, S. 35—36, 135, 274
Jehovah's Witness 14, 33, 135, 200
Jonsen, A. 71—175

Kant, I. 63
Katz, J. 3, 107—108, 270
Knight, J. 3—28, 85, 88, 140, 197, 230, 246

Macklin, R. 14—15
McCullough, L. 31, 38—40, 97, 112, 270, 275—276

Medicare 223
Meisel, A. 25, 254—255, 265
mental retardation 21, 85
mental status examination 10, 21
mens rea 77—78
Mill, J. S. 11
Miller, B. 20, 96, 131, 244
Morreim, E. H. 93—125, 199, 230

National Commission for the Protection of Human Subjects 75
Nozick, R. 232

organic brain impairment 20, 207
outcome 17—18, 131, 157—158, 255, 263

Parsons, T. 5
paternalism 39, 189—190
Pellegrino, E. 29—45, 198, 230
Perl, M. 179—184, 187, 190
Pincoffs, E. 79—87, 230
Pittsburgh Group 65—66
preferences 62
President's Commission xv, 17, 25, 83—84, 88, 138, 220, 256—257
proxies 40—41
psychiatry 3—4, 6—15
psychosis 20, 209

rational decision-making 10, 18, 30, 65, 96—98, 156, 180, 255
rational observer 100, 180, 241—242
rationality 262—263
Rawls, J. 63
reasons 18, 30, 34, 96—98, 180, 256
risk/benefit ratio 10, 19, 23, 65, 263
Robertson, J. 119, 127—148, 198, 228—230, 246
Roth, L. 17, 25, 131, 135, 179, 180, 254—5, 270

Schaffner, K. 253—284
schizophrenia 171—174, 190—191
Shelp, E. E. 181, 185—194
Sherlock, R. 257—258
Siegler, M. 36, 38, 71, 257—258, 264

surrogates 40—41
Szasz, T. 216

temperaments 201—209
treatment refusal 150, 168—169

understanding 18—19, 150, 180, 255, 262, 267

Vanderpool, H. 197—210

volition 187, 258, 264—265, 266—274
voluntariness 256—257, 267

Wear, S. 227—236
welfare 246
White P. 149—163
Wikler, D. 31
Winslade, W. 71

Youngner, S. 36, 135

Philosophy and Medicine

1. H. Tristram Engelhardt, Jr. and S.F. Spicker (eds.): *Evaluation and Explanation in the Biomedical Sciences.* 1975   ISBN 90-277-0553-4
2. S.F. Spicker and H. Tristram Engelhardt, Jr. (eds.): *Philosophical Dimensions of the Neuro-Medical Sciences.* 1976   ISBN 90-277-0672-7
3. S.F. Spicker and H. Tristram Engelhardt, Jr. (eds.): *Philosophical Medical Ethics: Its Nature and Significance.* 1977   ISBN 90-277-0772-3
4. H. Tristram Engelhardt, Jr. and S.F. Spicker (eds.): *Mental Health: Philosophical Perspectives.* 1978   ISBN 90-277-0828-2
5. B.A. Brody and H. Tristram Engelhardt, Jr. (eds.): *Mental Illness.* Law and Public Policy. 1980   ISBN 90-277-1057-0
6. H. Tristram Engelhardt, Jr., S.F. Spicker and B. Towers (eds.): *Clinical Judgment: A Critical Appraisal.* 1979   ISBN 90-277-0952-1
7. S.F. Spicker (ed.): *Organism, Medicine, and Metaphysics.* Essays in Honor of Hans Jonas on His 75th Birthday. 1978   ISBN 90-277-0823-1
8. E.E. Shelp (ed.): *Justice and Health Care.* 1981
     ISBN 90-277-1207-7; Pb 90-277-1251-4
9. S.F. Spicker, J.M. Healey, Jr. and H. Tristram Engelhardt, Jr. (eds.): *The Law-Medicine Relation: A Philosophical Exploration.* 1981   ISBN 90-277-1217-4
10. W.B. Bondeson, H. Tristram Engelhardt, Jr., S.F. Spicker and J.M. White, Jr. (eds.): *New Knowledge in the Biomedical Sciences.* Some Moral Implications of Its Acquisition, Possession, and Use. 1982   ISBN 90-277-1319-7
11. E.E. Shelp (ed.): *Beneficence and Health Care.* 1982   ISBN 90-277-1377-4
12. G.J. Agich (ed.): *Responsibility in Health Care.* 1982   ISBN 90-277-1417-7
13. W.B. Bondeson, H. Tristram Engelhardt, Jr., S.F. Spicker and D.H. Winship: *Abortion and the Status of the Fetus.* 2nd printing, 1984   ISBN 90-277-1493-2
14. E.E. Shelp (ed.): *The Clinical Encounter.* The Moral Fabric of the Patient-Physician Relationship. 1983   ISBN 90-277-1593-9
15. L. Kopelman and J.C. Moskop (eds.): *Ethics and Mental Retardation.* 1984
     ISBN 90-277-1630-7
16. L. Nordenfelt and B.I.B. Lindahl (eds.): *Health, Disease, and Causal Explanations in Medicine.* 1984   ISBN 90-277-1660-9
17. E.E. Shelp (ed.): *Virtue and Medicine.* Explorations in the Character of Medicine. 1985   ISBN 90-277-1808-3
18. P. Carrick: *Medical Ethics in Antiquity.* Philosophical Perspectives on Abortion and Euthanasia. 1985   ISBN 90-277-1825-3; Pb 90-277-1915-2
19. J.C. Moskop and L. Kopelman (eds.): *Ethics and Critical Care Medicine.* 1985
     ISBN 90-277-1820-2
20. E.E. Shelp (ed.): *Theology and Bioethics.* Exploring the Foundations and Frontiers. 1985   ISBN 90-277-1857-1
21. G.J. Agich and C.E. Begley (eds.): *The Price of Health.* 1986
     ISBN 90-277-2285-4
22. E.E. Shelp (ed.): *Sexuality and Medicine.*
    Vol. I: Conceptual Roots. 1987   ISBN 90-277-2290-0; Pb 90-277-2386-9

Philosophy and Medicine

23. E.E. Shelp (ed.): *Sexuality and Medicine.*
 Vol. II: Ethical Viewpoints in Transition. 1987
 ISBN 1-55608-013-1; Pb 1-55608-016-6
24. R.C. McMillan, H. Tristram Engelhardt, Jr., and S.F. Spicker (eds.): *Euthanasia and the Newborn.* Conflicts Regarding Saving Lives. 1987
 ISBN 90-277-2299-4; Pb 1-55608-039-5
25. S.F. Spicker, S.R. Ingman and I.R. Lawson (eds.): *Ethical Dimensions of Geriatric Care.* Value Conflicts for the 21th Century. 1987
 ISBN 1-55608-027-1
26. L. Nordenfelt: *On the Nature of Health.* An Action- Theoretic Approach. 1987
 ISBN 1-55608-032-8
27. S.F. Spicker, W.B. Bondeson and H. Tristram Engelhardt, Jr. (eds.): *The Contraceptive Ethos.* Reproductive Rights and Responsibilities. 1987
 ISBN 1-55608-035-2
28. S.F. Spicker, I. Alon, A. de Vries and H. Tristram Engelhardt, Jr. (eds.): *The Use of Human Beings in Research.* With Special Reference to Clinical Trials. 1988 ISBN 1-55608-043-3
29. N.M.P. King, L.R. Churchill and A.W. Cross (eds.): *The Physician as Captain of the Ship.* A Critical Reappraisal. 1988 ISBN 1-55608-044-1
30. H.-M. Sass and R.U. Massey (eds.): *Health Care Systems.* Moral Conflicts in European and American Public Policy. 1988 ISBN 1-55608-045-X
31. R.M. Zaner (ed.): *Death: Beyond Whole-Brain Criteria.* 1988
 ISBN 1-55608-053-0
32. B.A. Brody (ed.): *Moral Theory and Moral Judgments in Medical Ethics.* 1988
 ISBN 1-55608-060-3
33. L.M. Kopelman and J.C. Moskop (eds.): *Children and Health Care.* Moral and Social Issues. 1989 ISBN 1-55608-078-6
34. E.D. Pellegrino, J.P. Langan and J. Collins Harvey (eds.): *Catholic Perspectives on Medical Morals.* Foundational Issues. 1989 ISBN 1-55608-083-2
35. B.A. Brody (ed.): *Suicide and Euthanasia.* Historical and Contemporary Themes. 1989 ISBN 0-7923-0106-4
36. H.A.M.J. ten Have, G.K. Kimsma and S.F. Spicker (eds.): *The Growth of Medical Knowledge.* 1990 ISBN 0-7923-0736-4
37. I. Löwy (ed.): *The Polish School of Philosophy of Medicine.* From Tytus Chałubiński (1820–1889) to Ludwik Fleck (1896–1961). 1990
 ISBN 0-7923-0958-8
38. T.J. Bole III and W.B. Bondeson: *Rights to Health Care.* 1991
 ISBN 0-7923-1137-X
39. M.A.G. Cutter and E.E. Shelp (eds.): *Competency.* A Study of Informal Competency Determinations in Primary Care. 1991
 ISBN 0-7923-1304-6